Performance Interventions
Series Editors: **Elaine Aston**, University of Lancaster, and **Bryan Reynolds**, University of California, Irvine

Performance Interventions is a series of monographs and essay collections on theatre, performance, and visual culture that share an underlying commitment to the radical and political potential of the arts in our contemporary moment, or give consideration to performance and to visual culture from the past deemed crucial to a social and political present. *Performance Interventions* moves transversally across artistic and ideological boundaries to publish work that promotes dialogue between practitioners and academics, and interactions between performance communities, educational institutions, and academic disciplines.

Titles include:
Alan Ackerman and Martin Puchner (*editors*)
AGAINST THEATRE
Creative Destructions on the Modernist Stage

Elaine Aston and Geraldine Harris (*editors*)
FEMINIST FUTURES?
Theatre, Performance, Theory

Maaike Bleeker
VISUALITY IN THE THEATRE
The Locus of Looking

James Frieze
NAMING THEATRE
Demonstrative Diagnosis in Performance

Lynette Goddard
STAGING BLACK FEMINISMS
Identity, Politics, Performance

Alison Forsyth and Chris Megson (*editors*)
GET REAL: DOCUMENTARY THEATRE PAST AND PRESENT

Leslie Hill and Helen Paris (*editors*)
PERFORMANCE AND PLACE

D.J. Hopkins, Shelley Orr and Kim Solga
PERFORMANCE AND THE CITY

Amelia Howe Kritzer
POLITICAL THEATRE IN POST-THATCHER BRITAIN
New Writing: 1995–2005

Marcela Kostihová
SHAKESPEARE IN TRANSITION
Political Appropriations in the Postcommunist Czech Republic

Jon McKenzie, Heike Roms and C.J W.-L. Wee (*editors*)
CONTESTING PERFORMANCE
Emerging Sites of Research

Melissa Sihra (*editor*)
WOMEN IN IRISH DRAMA
A Century of Authorship and Representation

Performance Interventions
Series Standing Order ISBN 978–1–4039–4443–6 Hardback
978–1–4039–4444–3 Paperback
(*outside North America only*)

You can receive future titles in this series as they are published by placing a standing order. Please contact your bookseller or, in case of difficulty, write to us at the address below with your name and address, the title of the series and the ISBN quoted above.

Customer Services Department, Macmillan Distribution Ltd, Houndmills, Basingstoke, Hampshire RG21 6XS, England

Shakespeare in Transition

Political Appropriations in the Postcommunist Czech Republic

Marcela Kostihová

palgrave
macmillan

First published 2010 by
PALGRAVE MACMILLAN

Palgrave Macmillan in the UK is an imprint of Macmillan Publishers Limited, registered in England, company number 785998, of Houndmills, Basingstoke, Hampshire RG21 6XS.

Palgrave Macmillan in the US is a division of St Martin's Press LLC, 175 Fifth Avenue, New York, NY 10010.

Palgrave Macmillan is the global academic imprint of the above companies and has companies and representatives throughout the world.

Palgrave® and Macmillan® are registered trademarks in the United States, the United Kingdom, Europe and other countries.

ISBN 978–0–230–20324–2 hardback

This book is printed on paper suitable for recycling and made from fully managed and sustained forest sources. Logging, pulping and manufacturing processes are expected to conform to the environmental regulations of the country of origin.

A catalogue record for this book is available from the British Library.

Library of Congress Cataloging-in-Publication Data

Kostihová, Marcela, 1974–
Shakespeare in transition: political appropriations in the post-communist Czech Republic / by Marcela Kostihová.
p. cm. — (Performance interventions)
Includes bibliographical references and index.
ISBN 978–0–230–20324–2
1. Shakespeare, William, 1564–1616—Appreciation—Czech Republic. 2. Shakespeare, William, 1564–1616—Translations into Czech—History and criticism. 3. Politics and literature—Czech Republic—History—20th century. I. Title.
PR2971.C85K67 2010
822.3'3—dc22

2010027527

10 9 8 7 6 5 4 3 2 1
19 18 17 16 15 14 13 12 11 10

Printed and bound in Great Britain by
CPI Antony Rowe, Chippenham and Eastbourne

Contents

List of Figures

Acknowledgements

This project would not have been possible without the generosity of the Czech scholars, researchers, theatre directors, producers and actors who were willing to share with me their experiences with – and expertise in – postcommunist Shakespeare. I am thankful to Frantisek Čachotský, Andrea Cerqueirová, Michal Dlouhý, Vojta Dufit, Vilma Frantová, Gustav Hašek, Jan Helešic, Miloslav Hlaučo, Břetislav Hodek, Jiří Hromada, Anna Javorková, Jiří Josek, Jana Kavánková, František Kreuzmann, Zdeněk Kučera, David Matásek, Miroslav Mejzlík, Lenka Mrvová, Tomáš Pavlík, Lucie Peterková, Zuzana Pokorná, Jan Potměšil, David Prachař, Jan Přeučil, Daniel Přibyl, Ivan Rajmont, Michal Rychlý, Martin Sitta, Jitka Sloupová, Petra Špalková, Alena Štréblová, Vladimír Strnisko and Rostislav Trtík for their time and willingness to discuss their contributions to and views of postcommunist Czech theatre and Shakespeare's role therein. I would like to extend much gratitude to Michal Dočekal for extensive, enlightening and immensely informative conversations about his direction, artistic vision and the future of Czech theatre. Shakespeare scholars and translators Martin Hilský and Zdeněk Stříbrný have not only shared their work on Czech Shakespeare, but introduced me to many professionals in the field and generously offered to comment on early drafts of many of the following chapters.

Materially, this book has been supported by grants, at the early stages, from the University of Minnesota (Doctoral Dissertation Special Grant, Doctoral Dissertation Fellowship and the Samuel Holt Monk-Moses Marston Research Award) and finished with the support of a series of Hanna Grants from Hamline University. Several institutions and organizations have willingly allowed me to rummage through their theatre archives, namely Agentura SCHOCK, Divadlo Komedie, Národní Divadlo (Czech National Theatre) and Národní Knihovna (Czech National Library). Others have generously provided much-needed materials relevant to this project, such as Divadlo ABC, Divadlo Kašpar and Pidivadlo. I am particularly grateful to the nimble and generous help of the staff at Divadelní Ústav (Prague Theatre Institute) who assisted tirelessly with long hours of research, tracked trends in Shakespeare postcommunist performance, provided much-needed statistics on the postcommunist theatre-going public and, in one memorable case, ran their data to provide statistics for this book alone. I am grateful to the inspired photographers who spent countless hours documenting nuances of individual Czech Shakespeare performances, and whose insight frequently helped me broaden my views of each production in question. In this book, I use (with permission) the work of Tomáš Didunyk (production stills from *The Merchant of Venice* at Theatre Komedie), Viktor Krombauer (production stills from *Twelfth Night* at the Summer Shakespeare Festival), Ivo Mičkal (production stills from *Twelfth Night* at the Summer Shakespeare Festival), Iva Smejkalová (production stills from *Twelfth Night* at the National Theatre), and Milan Špelda (production stills from *The Taming of the Shrew* at Theatre Komedie). Several of the following chapters appeared – in preliminary iterations – in print. An early version of Chapter 3, 'Translation as Treason', has appeared in *Folio* (12, 2 [2005]); a short rumination on Chapter 4, 'Katharina Humanized', was published as a chapter in *Worldwide Shakespeares: Performance and Adaptation* (Routledge, 2005). A partial version of Chapter 5, 'Post-communist Nights', was included as a chapter in *Twelfth Night: Critical Essays* (Routledge, 2010).

As this book has been long in coming, there are many colleagues who need to be thanked for their generous insights on various iterations of the argument. The English Department at Hamline University could not have been kinder in supporting me through this project; I am lucky to be in such excellent company. Thomas Augst, Arianne Balizet, Melanie Brown, Andrew Elfenbein, Shirley Nelson Garner, Toni McNaron, Christina Schmid, David Slater, John Watkins and Allison Wee have provided substantive feedback – and moral support – in the early stages of this project. Particular thanks are due to the Medieval and Early Modern Research Group, namely Julie Eckerle, Karolyn Kinane, Linda Shenk and Jennifer Young, whose unwavering support, careful – and repeated – reading, exquisite intellectual challenges and warm friendship sustained me in many a tight spot. I am grateful to Juliette Cherbuliez for her prodding in discovering neoliberalism, Jean Howard for unfailing optimism, and Stephen Kellert and Mike Reynolds for sharp eyes and kind words. The series editors, Bryan Reynolds and Elaine Aston, saw the promise in this work before I saw it myself, no mean feat. The final version that is in front of you would not have materialized without the loving editorial ministrations of Jon Lloyd and the fabulous indexing prowess of Mikayla Moffet.

Last but not least, much gratitude is due to my families in the US, the Czech Republic and Germany. I would not have progressed an inch without my sister, who put in countless hours with the little people and shut me away in a room of my own. Děkuju, Kristýnko. Finally, I could not do without the Meyers: Sabine, Yasha and captain Voyta. It is to them that I dedicate this book.

Introduction

> *At this moment, I am still rather apprehensive about the fate of the perestroika and Mr. Gorbachev himself. But I am quite confident about the fate of Shakespeare in Eastern and East Central Europe. He will stay with us in our better or, if need be, our worse times.*
>
> Zdeněk Stříbrný, 1988 (2007: 211)

I begin with an anomaly.

In the early summer of 2001, I hurried through the crooked, impeccably maintained, tourist-infested streets of Old Prague to 'Montmartre', a small, recently opened theatre. Tucked away in a medieval cellar of a renaissance courtyard in a seemingly tentative alley off the main tourist routes, the theatre – like many similar venues scattered throughout the nooks and crannies of the city – caters to intimate groups of Czech theatrical clientele. As such, it symbolically preserves a corner of Czech cultural capital from the frenzy of well-orchestrated international consumption of Prague's 'authentic' historical spirit. The performance I was hurrying to see, however, was not of a new Czech play; that evening marked the last performance of the season of Shakespeare's *Sonnets*.

A performance of *The Sonnets*, in the Czech context, is not anomalous in itself. Though not staged frequently, the poems have made regular appearance on Czech stages since the nineteenth century. They have usually been offered to an appreciative public as performative readings by distinguished actors, sometimes accompanied by a small ensemble of period instruments that tastefully underscored the cultured mood of the event, sometimes by a relevant lecture on the poems, their context or the pitfalls of their translation. In keeping with the larger cultural context of ongoing commitment to Shakespeare, manifested in a steady, well-attended stream of productions of his plays, these evenings have drawn crowds of spectators and have been jubilantly reviewed by the press. The anomaly of this June evening was of a different, entirely unexpected kind: I was the only one in attendance. Instead of the usual press of well-dressed Czechs buying out the small venue, I found an obscure, shuttered window, long abandoned by all except for a few urbane pigeons. Puzzled, I inquired at the adjacent café about the whereabouts of its neighbours. The barista waved me in the direction of a corner table where I met Tomáš Pavlík, Rostislav Trtík and František Kreuzmann, the director and two actors scheduled – but not sought to – perform that night. Instead of watching *The Sonnets*, I ordered ale and conducted the first in a series of interviews about their experience of staging a Shakespeare classic.

Like all Czech dramatizations of the sonnet cycle, the Montmartre *Sonnets* were initially well-attended. The lustre of the actors was buttressed by consistent positive reviews of the trio's previous work, which experimented with dramatizing short stories, essays and, as in the case of *The Sonnets*, poetry. Their latest production, a dramatization of Tommaso Landolfi's short stories, was concurrently performed in the most prestigious of Prague's small intellectual venues, Lyra Pragensis. The *Sonnet* production further received ample encouragement from Martin Hilský, an eminent Shakespeare translator whose prize-winning *Sonnets* the trio used as their initial script, and from several other prominent Prague theatre professionals. Yet the production did not fare well with the general audience or the media. The number of attendees dwindled to single digits and professional critics, who had praised the trio's other work in Czech newspapers and journals, did not seem to find the production worth their publicized attention.

The production itself, which I saw several nights later at a special staging, took the performative aspects of Shakespeare's poetry beyond the usual genteel animated reading to a spirited enactment of a narrative – pieced together from individual sonnets – that explores intense emotional, erotic and sexual desire experienced by *The Sonnet*'s speaker and his masculine object of affection. Challenging the pervasive misapprehension of *The Sonnets* as a refined, venerable and cultured expression of normative heterosexual love, the two actors used Shakespeare's text as a platform to speak passionate desire made ambivalent by a lack of established homoerotic cultural discourse. Staged on a minimalist set depicting an urban landfill, the performance powerfully underscored the possibilities of raw human affection at the level of bare existence at the outskirts of a society that increasingly turns its gaze away from disenfranchised humanity to profit and consumerism, leaving those on the margins literally to rot in a dump.

The anomalous silent treatment dealt to the production by the otherwise easily-pleased audience and media arose – at least in part – from the explicit portrayal of homosexuality and homelessness. Even so, this response seems disproportional to the cultural intervention this performance attempts. After all, an explicit portrayal of homosexuality could not possibly create that much cultural havoc; around the same time, a prominent Prague theatre, Divadlo Karlín, profitably staged a sequence of US musicals with explicit homosexual themes, *The Birdcage* and *Victor-Victoria*. The throngs of spectators who streamed to see these high-profile shows suggested a contrary fascination with portrayals of non-normative sexuality on stage. Similarly, the spectacle of men in rags in postcommunist Czech theatre has not – to my knowledge – elicited unexpected negative responses from either the public or the media. Perhaps the critical combination of queers and/in rags, suggesting that somehow the postcommunist society pushes the former into the latter, could be identified as the most likely culprit for the negative reception. Yet, rather than publishing negative reviews, the Czech print media shrouded *The Sonnet* production in absolute silence. This wilful blind spot (which preempted further public dialogue about the subject matter) suggested an unusual implicit agreement among otherwise contentious media outlets to gloss over whatever 'message' the production might appear to disseminate.

Rather than being a result of the mere public portrayal of same-sex desire – or of homelessness for that matter – the anomalous response to the production seems to originate from the anomalous appearance of these subjects *in Shakespeare*. Whereas Czechized versions of American musicals provide glimpses of pleasantly foreign non-normativity presumably untranslatable into the world of everyday Czech culture, performances of Shakespeare speak to a complex articulation of Czech perception of universal humanity, which is central to the increasingly conflicted postcommunist reconstitutions of Czech nationhood within the ideological premises of Western democratic selfhood. Such evocation builds on a long-term, fervent commitment to Shakespeare as a model of transcendental human essence, a model that has served as a tool of national preservation as well as of intellectual and cultural resistance to relatively rapidly changing colonial regimes of the past two centuries. Similarly, in the midst of the transitional turmoil, postcommunist Czech Shakespeare productions have been reflecting – and participating in – ongoing public, occasionally-conflicted discourse of the country's transitional process, a discourse often divorced from the official political negotiations of the Czech government with its neighbors in regards to Czech integration into the European Union (EU). Amid the growing scepticism of the integration process in the period this book addresses (1995–2005), in which the surface of the EU's lofty rhetoric of equality revealed a persistent practice of exploitive neoliberal neo-imperialism that has significantly eroded the living conditions of the majority of average Czech citizens, Shakespeare productions have begun reclaiming the space of potentially subversive political discourse.

The Montmartre *Sonnets* performance stepped into a turbulence of converging cultural and political torrents that were exacting powerful reactions from the public and the media. The icy reaction to explicit portrayal of homosexuality in Shakespeare harnesses an established Czech xenophobia to buttress a much larger – and far more complex – cultural insistence on the right to postcommunist development free from interference from yet another set of foreign powers. Even as the production was taking shape, the Czech government was feverishly seeking to enact legislation that would satisfy the non-negotiable conditions – the first such conditions ever stipulated – set by Western European countries for the accession of the Czech Republic and several other candidates from Central and Eastern Europe (CEE) to the EU. Aiming to provide evidence of the nation's 'respect and protection of minorities', the parliament debated ways to include the 'homosexual minority' in legislation guiding the domestic and public spheres (for example, domestic partnership/marriage, adoption, workplace discrimination, etc.). On the one hand, such discussions and subsequent legislation undeniably signalled positive developments in the contested area of human rights in CEE in bringing higher visibility and material enfranchisement to a hitherto repressed, pathologized – or at least systematically ignored – minority of Czech citizens. On the other hand, the enfranchising process itself – particularly as it evoked the question of national and European citizenship and belonging – uncovered startling inconsistencies in international rhetoric that sought to shape the future of the Czech Republic and its CEE neighbours. These tensions that surrounded the gradual articulation of postcommunist normative subjectivity projected into all areas of social life, surfacing particularly in cultural spaces – such as Shakespeare production – that have been traditionally intimately linked to the symbolic core of Czech cultural and national identity.

Far from being parochial in their reach, Czech reservations about the required inclusion of homosexuality within the normative boundaries of postcommunist Czech subjectivity, which the Montmartre production inadvertently addresses, exemplify international concerns about national agency in determining the dimensions of normative humanity and relevant citizen rights within the context of the powerful EU. The international political dimensions of the inherent tensions unearth the sometimes paradoxical intersections of the EU social, civil and economic accession requirements. The provision for 'the respect and protection of minorities', for one, is included in the same list of conditions for admission into the EU as the bitter neoliberal shock therapy pills of free markets that, since 1990 when they were first implemented on the 'advice' of international experts, systematically diminished the living standards and securities of most Czech citizens. In the simplest sense, any rhetoric of human rights, however well-intentioned, seems suspect in the face of the material results of EU policies that effectively and systematically *disenfranchise* the majority of citizens through enforced layoffs in the name of 'flexibility' and 'efficiency' of the labour force, inequality in applying EU subsidies for key economic sectors (such as the highly-contested agricultural sector), discriminatory application of nominally universal rights to seek employment internationally within the EU, and diminishing social security provisions, while wealth evaporates upwards towards a small wealthy elite and/or international corporations dispatched by Western governments ostensibly to assist with the transitional process.

The backlash against widening the normative parameters of postcommunist subjectivity stems from a wide cultural disappointment with the transitional process, particularly in the failure of expected democratic freedom. The focus on 'freedom' and 'liberation' from totalitarian control, central to the revolutionary events of 1989, has been initially subsumed into the neoliberal frameworks of individual 'private freedoms' most concerned with state protection of private property and means of consumption rather than with systemic support of egalitarian distribution of basic material resources within the framework of social support. The rapid privatization of the public sphere, in the guise of the restitution of private property nationalized by the communist government after 1948 and of postcommunist privatization of key state enterprises, has amplified the inequalities of access to the material frameworks that, quite pragmatically, make civil liberties meaningful. The parallel rhetoric of democracy, often welded by neoliberalism to free markets, is belied by the lack of democratic process that marks much of the implementation of postcommunist, EU-oriented policies. While the rhetoric of democracy has been a centrepiece of all deliberation about the accession to the Union, the form as well as the process of implementing such a framework has proved paradoxically undemocratic. The accession conditions, dictated by Western members who would most likely not satisfy them if examined,[1] are non-negotiable; this inflexibility by definition excludes substantive input by those whose future such conditions will chart. Moreover, despite nominally assumed equality between existing and potential members of the EU that would assure a relative uniformity in freedom applied to building democracy in each EU nation state, the reality that the conditions applied to potential CEE members are distinctly neoliberal in character – sharply diverging from the far more socially protectionist democracies of Western Europe – undermines the democracy-building process at its very outset.

Inevitably, the postcommunist redefinition of the Czech nation is haunted by a rising paradox of the Union: while membership of the Union assumes a degree of readiness for European post-national citizenship, championed by the federalist faction of EU thinkers, the entire process is built – and enforced – along nationalistic lines. Within the EU-supervised ongoing process towards meeting the necessary targets for unionization, rules are stipulated and privileges are bestowed on groups of *nationals*, so that Czech economic policy, for instance, determines the degree of Czech citizens' cross-border employment mobility. While the over-arching rhetoric of the EU seems to promote European identity, its policies reinforce the very identification that its rhetoric seems to work to dismantle. The wishful-thinking dimensions of the unionist rhetoric has further been belied by intensifying conservative nationalism in Western EU member states in response to fears of negative material impacts of expected and actual immigration from the CEE.

It is perhaps not surprising, then, that the focus on the nation, particularly as the locale of newly rising resistance to collective nation-defined disenfranchisement, has become a centrepiece of post-communist restructuring of the Czech state. It is equally unsurprising that the early postcommunist Czech enthusiasm about joining the Western world has given way to a measure of disillusionment and scepticism with the transitional process, paralleled with strategic deceleration of early eager neoliberal moves in favour of cautious social protectionism and suspicion towards all suggestions by the EU. While the nation could not existentially afford to boycott the process of integration into the EU, which meant that the stipulated non-negotiable conditions would be met, the strategic acquiescence of government officials has not been mirrored by a corresponding shift in social relations. Indeed, thanks to four centuries of colonial domination by three distinct empires, the Czechs have gathered extensive experience in subversive resistance well-covered with surface compliance. Václav Havel's observation in *The Power of the Powerless* that totalitarianism leads to 'living a lie' (1985: 31), which Katherine Verdery has noted as strategic 'internal sabotage' of the system (1996: 23), speaks to a duality that had become habitual to generations of citizens. Far from being duplicitous, in the interests of cultural and national preservation in the face of pervasive external domination, resistant subjects have developed circumstantially adaptable – and thus surprisingly enduring – collective frameworks of cohesive cultural resistance. Conveniently for this project, Shakespeare has persistently functioned as one of the stable mainstays of this mass resistance, providing a nominally universal cultural capital that transcends whatever temporal limitations the Czech nation experiences. What could or would not be asserted in the public political sphere could be 'found' in Shakespeare's genius work and staged accordingly. Despite claims that, in a postcommunist free democracy, Shakespeare no longer engenders political interpretations, the close of the twentieth century marked a progressive increase in the transitional public sphere projected into Shakespeare productions.

The Montmartre *Sonnets* – by advocating the rights of minorities via the cultural capital of transcendental Shakespeare – came perilously close to seeming collaborationist with the culturally threatening European integration conditions. The media silence – and the silent boycott of the audience – could be recast as tentative beginnings of new cultural resistance of a forming double-consciousness split between necessary outward political changes on the one hand and, on the other hand, a solidifying collective imaginary national core. In this early stage, subtlety seemed unwelcome, for the collective response clearly missed the production's critical stance that potentially exposed the fallacy of the liberationist rhetoric within the postcommunist transition. After all, while the two characters of Shakespeare's *Sonnets* were apparently free to experiment with expressing hitherto forbidden affection for each other, their 'freedom' came at the steep price of spatial marginalization in the garbage dump of consumerist capitalism. If, the production seems to suggest, homosexuals are part of transcendentally-identified universal humanity, in the postcommunist context, that humanity tumbles down the economic ladder to consume freely the unwanted leftovers of the wealthy.

The anomaly of the Montmartre production provides a fitting entry into an inquiry into the intersections of postcommunist culture with its politics. These intersections sometimes painfully reveal the sociopolitical undercurrents that influence the transitional process of a precariously positioned postcommunist nation. In the following pages, I will attempt to illustrate the ways in which the official political process of postcommunist 'Westernization' does not fully account for the directions this transition may take. On the contrary, in considering the ethnographic perspective of the transition, I zero in on the moments where a new current of subversive dissent begins to cohere in the face of profound uncertainty and perceived threats arising from the transitional process hatched in the midst of a long history of colonial oversight, as well as an unavoidable future within another powerful neo-imperial entity. In watching this transition unfold in its multiple layers of creation, negotiation and coercion, we can study the formation and meaning of Western cultural practices themselves.

Why the Czech Republic and why now? By limiting my inquiry to the decade between 1995 and 2005, I address a rich moment of transitional development book-ended by, on the one side, the waning exuberance of postcommunist freedom of the mid-1990s and, on the other, the accession to the EU in 2004. In this transitional window, Czech society struggled to piece together its own cultural identity after four centuries of repeated physical and ideological colonization even as it negotiated the tensions arising from semi-voluntary unification with another neo-imperial entity. This struggle to assemble the pieces into a workable whole was similarly undermined by the rhetorical experiments in dismantling national frameworks of identification in favour of a new, pan-European, constitution-based identity, a rhetoric immediately undermined by laws, limitations and conditions applied on the basis of the nation-state and thus the nationality of each affected citizen. This period was further marked by the globe-shaking ideological, political and material consequences of the terrorist attacks of September 11 followed by a Western invasion of Afghanistan and Iraq that further exposed the increased fracturing in the façade of Westernism.

The geopolitical site of the Czech Republic enables a rewarding study of postcommunist Shakespeare, not only because of the profound ways in which the Czechs have made Shakespeare Czech – an appropriation paralleled in similar ways in other postcolonial and postcommunist nations, as I will explore in Chapter 2 – but primarily because of the position that the Czech Republic has claimed on the political map of Europe. Literally in the centre of the continent, it is sandwiched between traditionally Western (Anglo-Saxon and Romance) and traditionally Eastern (Slavic) cultures. In the endless competition of empires, it has been claimed by a series of colonizers who have sought to annex its territory, natural resources and strategic locale (a competition that continues, at present, with the escalating international tensions surrounding US plans for an anti-missile defence shield facilities on Czech territory). Despite – or, I begin to argue, because of – this repeated domination, the Czech culture has developed unique means of self-identification projected emphatically into the cultural sphere. The Czech appropriation of Shakespeare, then, has been driven by a complex set of unique forces which, when investigated, yield an intriguingly novel consideration of literature, politics and nationhood.

Throughout this book, I attempt to bring together the wider context of postcommunist sociopolitical and economic developments, their impact on the quality of everyday experience of average citizens, and the sphere of the presumably depoliticized Shakespeare. I am particularly drawn to the nature of the cultural capital that Shakespeare uneasily represents; while nominally equated with the idea of a non-ideological universal humanity, it simultaneously embodies Westernism *and* the core of the postcommunist, inevitably national culture. In the initial stages of postcommunist development, when a 'return' to Western 'Europe' seemed possible, these three ideological discourses were believed to be unified. Subsequently, however, the West-imposed process of neoliberal shock therapy – with all attendant material results – made visible the already-existing ideological and material faultlines.

Methodologically, I rely on a multiplicity of sources and scholarly approaches. Inevitably, I draw in part on traditional archival research of Shakespeare scholarship and performance review, as well as news media, cultural studies, sociological data, economic statistics and political-science research that buttress my predominantly cultural-materialist theoretical framework. I have seen either in person or on archival video all the productions which I discuss at any length. My production analysis takes into consideration the fact that Czech theatre operates predominantly on the repertory model and, largely because of decades of periodic visits from ideological censors, relies on a tradition of flexibility and improvisation that enables both the evolution of a performance over consecutive seasons as well as minute adjustments to the latest political developments and an audience's immediate mood. I have therefore made an effort to attend productions repeatedly and over time whenever possible to note the overall tenor and ideological thrust of each artistic endeavour, so that my subsequent analysis does not reflect impressions from a single performance of an otherwise multi-faceted production. When using video archived either by the Theatre Institute in Prague or by individual theatres (or production directors) for staffing/educational purposes, I have made the assumption (which I have verified whenever possible with theatre professionals involved in its creation) that theatre ensembles archived performances they deemed representative of the production as a whole, rather than random televised events.

In their interventions into the cultural sphere that frequently produce currents cutting into/under official Czech national and international policy, postcommunist productions of Shakespeare arise from the sometimes contradictory efforts of those who create them. In seeking to understand both the cultural interventions and the contradictions that create them, I have conducted a series of interviews with many of the artists and professionals – directors, producers, translators, scholars, photographers and principal cast members – involved with each production I address in the following chapters. Rather than inflexible monoliths of sharply articulated points of view, these interviews revealed that artists are responsive individuals who are as swept in the whirlwind of current affairs as their audiences, if perhaps more attuned to – or more intentional about reflecting on – their sociopolitical surroundings. In the course of repeated interviews, a few of which were conducted over several years between 2001 and 2007, it became clear that the views expressed by individual theatre professionals – particularly on the subject of socioeconomics and national politics – have evolved in response to the changes in the Czech social, economic, political and cultural climate as well as to personal experience and insight. All the excerpts from interviews used in the later chapters have been carefully vetted by my interview 'subjects', who in many cases graciously agreed to see in print views they had had formerly expressed but no longer espoused. Some interviews have been removed entirely. In any case, I have learned that all information gleaned from interviews – as tantalizingly close to 'the source' as it appears – is contextual and profoundly local, functioning not as rock-hard evidence of the role that Shakespeare has played in the postcommunist Czech Republic, but rather of the ways in which we all – and the cultural products we create – negotiate the often contradictory and individual currents that converge in each instance of cultural production.

Though Shakespeare as a form of cultural capital operates from a multiplicity of intertwined and cross-fertilizing institutional sites, I focus predominantly on exemplary instances of popular and media reception of particular Prague Shakespeare productions that surfaced in exceptionally charged moments of marked political challenge to the private sphere. Rather than seeking a comprehensive theory of Shakespeare use in the postcommunist process of cultural rearticulation, I am irresistibly drawn towards selective examples of Shakespeare negotiation of the tense confrontation between the internationally-oriented official public discourse and the rhetoric of individual agency and postcommunist subjectivity within the private sphere. The sudden multiplicity of emerging readings of Shakespeare that compete with each other on postcommunist stages and in the academic press, while nations scramble to rearticulate their relationship with the existing Western powers, speaks to the potential of the political resurgence of dissent to emerging sociopolitical frameworks as well as to an ongoing effort to articulate a viable postcommunist national identity.

I have divided the following book into two rough halves. Part I, which provides the theoretical framework as well as the sociopolitical, economic and cultural backdrop for the case studies that follow, is further subdivided into two chapters. Chapter 1, 'The "End of History" in Central and Eastern Europe', explores sociopolitical and economic developments in the region with a special focus on the socioeconomic impact of neoliberal postcommunist 'normalization' on the eve of accession to the EU. Juxtaposing the optimistic rhetoric of the EU's suggestions and requirements with socioeconomic data that shows a sharp decline in CEE living standards and an attendant weakening of traditional frameworks of identity formation, this chapter outlines the growing despair over postcommunist developments as well as growing Euroscepticism and distrust of the established West in the region.

Chapter 2, 'Shakespeares of the Postcommunist World', follows with a brief summary of the uses of Shakespeare's works and cultural capital for political purposes (broadly defined) generally, and then specifically in the CEE. The chapter continues with an examination of the specific legacy of Czech political and artistic developments to which Shakespeare has been central. First, I consider two distinct historical moments within which Shakespeare has been evoked as a tool of nation building defined in resistance to colonial dominance: I juxtapose the Czech Shakespeare tercentenary in 1864, demonstratively celebrated to subvert the ideological hegemony of the Austro-Hungarian Empire, with the Shakespeare quatrocentenary, which enabled a rearticulation of Czech nationalistic resistance to Soviet dominance in 1964. In outlining the central dimensions of the use of Shakespeare for such political purposes, I then turn to the postcommunist moment to outline the dimensions of the proliferating transitional Shakespeare.

Part II builds on the frameworks established in the first two chapters by offering particular case studies of clusters of Czech Shakespeare translations and/or performances arising in response to social tensions surrounding the cultural implications of joining the EU. Chapter 3, 'Translation Wars: Redefining Shakespeare in the Postcommunist Czech Republic', analyses arguments that surfaced in a 2001 public debate between Czech translators in regards to the 'correct' translation of Shakespeare in which the 'wrong' translation was not only perceived as a failure of the Czech culture, but as a direct 'treason' against it. I argue that these 'translation wars' were not a result of a simple disagreement about textual translation of Renaissance drama, or even of an economic dispute over whose translations should be used in order to make theatrical productions most lucrative. Rather, they stem from a profound instability surrounding the redefinition of the Czech cultural identity in the face of neoliberal developments that undermine emerging democratic frameworks and structures of individual belonging.

Chapter 4, 'Katharina "Humanized": Abusing the "Shrew" on Prague Stages', connects three 2001 Prague productions of *The Taming of the Shrew* within the context of the redefinition of gender and domestic abuse in the wake of the 1989 political changes. Drawing on the performances themselves, interviews with directors and actors involved in the productions, popular and scholarly reviews, and recent data on domestic abuse legislation required by the EU, I link the play's interpersonal violence with the crisis of traditional masculinity resulting from the material practices of neoliberal shock therapy. Beyond reflecting commonplace domestic violence in the culture at large, the productions – as well as the dominant popular responses to them – exemplify a general cultural insistence on preserving the autonomous privacy of the domestic sphere as a site of resistance to the widely suspected neoliberal rhetoric of human rights championed by the EU as part of the Czech accession process.

Chapter 5, 'Politics of Desire: Postcommunist Czech Shakespeare and Non-normative Masculinity', further develops the analysis of the masculinity crisis in discussing the (homo)sexual tensions in current productions of *The Merchant of Venice, The Sonnets* and *Twelfth Night*. Exploring the position of non-normative masculinities – and homosexuality as one of its salient examples – in the Czech culture through the contrast between the implicit 'high' and 'low' Czech art, the chapter compares the favourable reception of current productions of US musicals such as *The Birdcage* and *Victor-Victoria* with the hesitation with which such themes are explored in current Shakespeare productions. My analysis uncovers a cultural ambivalence about accepting homosexuality as a social reality, rather than as a fleeting, fashionable, Western-imported, EU-enforced and hence discountable trend.

Part I

1
'The End of History' in Central and Eastern Europe

Since it is my contention in this volume that postcommunist Czech Shakespeare has begun to serve as a newly emerging site of resistance to neoliberal practices of the neo-imperial West, particularly the European Union (EU), it is crucial to outline such policies – and their material consequences on Central and Eastern Europe (CEE) citizen subjects – at some length. It is only through the exploration of some of the most relevant examples of Western intervention in the postcommunist transitional process that we can begin to appreciate the dimensions of the emerging disbelief, sense of betrayal and subsequent mobilizing of cultural resistance to exploitive policies and cultural frameworks that enable them. I hope that my attention to the sociopolitical context here will provide an ample backdrop not only to my initial argument made in the Introduction, but also for the detailed consideration of particular incarnations of Shakespeare in the following chapters.

The postcommunist transition of the Czech Republic, though in many ways inevitably unique to the specific Czech context, cannot be considered in isolation from the rest of its CEE neighbours who found themselves in a similar predicament after 1989. In fact, much of the transitional process was driven by, on the one hand, an uneasy cooperation with other states in the region and, on the other hand, undoubted competition for Western funds, connections, economic agreements and inclusion in powerful Western organizations such as NATO and the EU. The driving constant, however, throughout the majority of CEE transitions has been a desire to create maximum political, economic and cultural distance between the postcommunist countries and the latest empire to control them, the Soviet Union.

In practice, this process of distancing CEE from the USSR translated into policies and processes that sought to liberate postcommunist systems from any guise of communist ideology, practice and state control. In its place, CEE countries sought to articulate models that would bring them 'back' into the protective fold of Western Europe that would ensure a future where Soviet – and later Russian – domination would not be likely to recur. In the early celebratory stage of the post-1989 transition, models of postcommunist sociopolitical organization proliferated and clear consensus – apart from the common denominator of anti-communism – was difficult to find. Not surprisingly, many CEE societies looked to the well-established rhetoric of anti-communist dissent that allegedly contained freedom-seeking alternatives to the repressive Soviet imperial structures. The philosophical basis of these models proved vague, if generally appealing in its rhetoric of rights and freedoms. The main thrust of anti-communist dissent had been to strive for a society that lives, as Václav Havel had coined the phrase, 'in truth'. Yet, while to 'live in truth' became the rallying cry of the recently liberated CEE citizens in the final months of 1989, the content of such 'truth' remained painfully unclear beyond the opposition to whatever falsehood was daily produced by previous colonial regimes. While in contexts befuddled by communist ideology and doublespeak such philosophy seemed to offer welcome clarity, it proved woefully inadequate as a solid basis for postcommunist society.

Yet, despite the surprising variety of capitalist models offered by the 'first world', CEE countries tended to fall into a duality of reactionism against the communist imperial ideology. As Jerzy Szacki has observed, this has meant that 'dogmatic Marxism [was] replaced by dogmatic liberalism' enabled by a practice of equating all dissenters of the communist regime with 'liberals', i.e. those who would bring freedom: 'it sufficed to be dissatisfied with the status quo and pine for some, even the most vaguely defined, liberty, or "greater liberty"' (Szacki, 1996: 25). Within this oppositional binary, the new system remained vaguely pivoted on the concept of 'freedom', defined around a resistance to a state apparatus that had symbolized an imposed foreign – and by then thoroughly discredited – socioeconomic ideology, a rhetoric of false Orwellian equality (where all were equal, but some were more equal than others), a totalitarian control of expression and a material standard of living that began stagnating and, in some areas, decreasing prior to 1989.

Such vague resistance already resonated with the rhetoric put forth by proponents of neoliberal practices quickly imported by economic and political experts dispatched by Western governments to assist with the transition in the midst of the postcommunist disorientation. Nominally organized around the concepts of individual freedom, democracy, deregulated economic enterprise and minimalization of state interference in private affairs, neoliberalism seemed to offer common sense steps towards the realignment of CEE countries with the 'free' Western world. Indeed, the greatest promise of neoliberalist experts was to transform CEE postcommunist nations into 'normal countries', that is, socioeconomic systems similar to developed Western Europe.[1]

The curative metaphors of medicine, bitter pills and surgery so often used by Western economic experts when referring to the proposed structural adjustments within CEE took an early hold. Taking advantage of the disaster-like conditions created by the precipitous fall of communist regimes in CEE, such advisers proposed drastic economic changes rooted, most importantly, in the privatization of state and collective industries (including social services), downsizing and deregulation of the labour force, and stark deregulation of the market. As Naomi Klein argues in her representative analysis of the Polish transition, 'the disorientation of rapid political change combined with the collective fear generated by an economic meltdown [made] the promise of a quick and magical cure – however illusory – too seductive to turn down' (2007: 181). Accordingly, and to various degrees, CEE governments embarked on a general restructuring of their economic systems, hoping for short-term growing pains leading to long-term gains that would rapidly contribute to their countries' stabilization and cement a future equitable position within the club of developed countries.

The restructuring process that was to reunite the forcefully estranged geographical ends of Europe to the benefit of the continent was to be partially funded by aid from the developed world. Multiple Western governments, including the USA, took advantage of the fall of communism in 1989 to make celebratory proclamations of the 'end of history', pledging funds and assistance in the process of unifying the newly 'liberated' countries with the developed 'first' world. Yet the follow-up proved reluctant (Trenin, 2003: 1; Klein, 2007: 176ff). Moreover, contrary to CEE expectations, in place of substantive relief of the debt incurred by communist governments after the Second World War and of financial aid, the bulk of the aid that materialized came in the guise of so-called 'technical assistance' (Wedel, 1998) or 'transition industry' (Swain, 2006: 208–10). This assistance predominantly consisted of the presence of Western 'experts' (international financial institutions, such as the IMF or the World Bank; professional economic consultants, sometimes known as 'econolobbyists'; and academics) who were to steer and monitor compliance to shock therapy guidelines as well as to establish ideologically-charged training programmes designed to ensure adherence to neoliberal models (Wedel, 1998: 30). In fact, despite Western rhetoric of a 'second Marshall Plan' that would enable a sustainable restructuring of CEE socioeconomic systems, the comparison disintegrated in the disparity of dedicated funds.[2] Instead of material aid, Wedel concludes, CEE countries received the 'Marshall Plan of Advice' (ibid.: 30). A large proportion of the dedicated finances, rather than being distributed for transitional CEE projects, made an elegant U-turn back West to pay the consulting fees of the Western advisers.

Besides the disparity in the actual amounts of distributed aid, this second 'Marshall Plan' was charted along startlingly divergent ideological lines from the first. The first US Marshall Plan provided financial aid to war-ravaged Western Europe to prevent the possibility of the spread of communism westwards (Klein, 2007: 252). Besides this general directive, the Plan was flexible in application, being designed to enable individual governments to restructure state economies according to principles democratically determined by each country's specific needs, traditions and existing situation, thus resulting in the vibrant diversity of predominantly social-democratic capitalist systems in the region. In contrast, in the absence of a similar communist threat after 1989, Western aid did not fear a popular revolt of disempowered labour in distributing transition funds. Accordingly, most of the post-communist aid was provided through established neoliberal channels so that available loans from international financial institutions were harnessed to economic conditions of 'structural adjustment' which generally consisted of the creation of 'free markets' operating in a 'modernized' economy of greater 'efficiency'. Substantively different from controlled and subsidized markets of the established West, this adjustment has allowed Western investors to treat CEE territories as a new frontier for astronomical Western profit or, in genteel parlance, a favourable climate for foreign investment: few and low taxes, a flexible and unorganized workforce, little oversight by the 'host' countries and a lack of obstructions to an extraditious flow of capital back to investors' pockets. Though claiming generously to assist CEE citizens – after all, as Katherine Verdery has noted, the 'shock therapy' represents Western advisers as doctors [and] the "big bang" figures them as God' (1996: 205) – the Marshall Plan of Advice was designed to primarily benefit the West, whether in the form of financial profits from CEE investments, expanding markets for Western goods or increased control of strategic territories of geopolitical significance.

As early as 1992, representatives of CEE countries included in the early wave of Western postcommunist 'assistance' began articulating their sense of betrayal and disenchantment at Western practices that funnelled much-needed resources into the pockets of private consultants at the expense of the CEE economies, 'draining [their] domestic markets' (Wedel, 1998: 30). Their disenchantment further stemmed from the reality that CEE governments had little or no input into how the aid – particularly the aid that originated in the US and to a lesser extent aid from the EU – for the 'recovery' of their constituent countries would be delivered and distributed; their exclusion from the decision-making process often translated into exclusion from the flow of information altogether (ibid.: 35).

The eventual outcomes of shock therapy were indeed radical, but not in a way that CEE governments had envisioned. Though these results have been well-mapped by economists, anthropologists, social and political scientists and, increasingly, scholars in the humanities, their salient highlights are worth repeating. In the region as a whole, living conditions sharply descended below those of the Great Depression of the late 1920s and 1930s. In 1997, well after shock therapy was promised to deliver miraculous results and long before the global economic crash of 2007 onwards that further compounded the socioeconomic devastation of the region, UNICEF found that 'the majority of peoples under transition suffer a deterioration of living standards, unemployment and poverty affecting every third citizen even in the most affluent countries' (reported by Moskalewicz, 2000: 563). The deteriorating quality and decreasing availability of social services was paralleled by skyrocketing unemployment, poverty, mortality, suicide rates, alcoholism, violence and a general widening of social inequalities (Štulhofer and Sandfort, 2005: 2). In an extensive study of 'transition economies', Branko Milanovic found that average real wages 'dropped by one-fourth between 1987–88 and 1994, while unemployment grew from zero percent to between 12 and 15 percent of the labor force' (Milanovic, 1998: 29). Historian Robin Okey additionally cites staggering regional drops in birthrates as an 'extreme form of belt-tightening'; the eradication of the middle class materially descending to near-poverty; youth unemployment; skyrocketing crime; slashes to healthcare funding to 10 per cent of EU levels in Poland, for instance; a crisis in the drug supply; and a decrease in cultural consumption and research (Okey, 2004: 124). Heavy industries and agriculture, being prime targets for privatization, were intensely affected by the downsizing and dislocation of the labour force, creating a further binary opposition between relatively affluent, service-industry oriented large urban areas and rural agricultural regions, as well as satellite cities built around individual, now faltering or eradicated industries. In general, income disparities that elevated a fraction of wage-earners towards unprecedented wealth while leaving the majority in a state of poverty or near-poverty were perhaps the most psychologically damaging in a region long-committed to an egalitarian *modus operandi*.

As the process of transition trickled to the level of the individual, most intimate precepts of individual identity were affected and, as a salient example, gender norms became increasingly unstable. The growing socioeconomic instability markedly created a widespread and widely-documented psychological crisis in the male population which created multi-dimensional tensions between generations, sexes, ethnicities and social classes. This 'male identity crisis' has closely paralleled a crisis of citizenship and belonging as they are defined in each respective culture, marking a significant interference in a sizeable population's ability to perform roles traditionally associated with appropriate citizenship. As Jacqui True has mapped out, the sudden economic disempowerment of shrinking real wages, unemployment, the rising costs of living and the rapidly decreasing social security networks were 'reflected in the rising male suicide and mortality rates, alcohol abuse, and unemployment and crime rates across the region' (2003: 71). Although a small proportion of younger, urban and educated men have been able to benefit from the neoliberal processes that tapped their potential to create a new class of business-oriented supporters, the majority of the male population has remained susceptible to continuing instability that has further exacerbated already-escalating frictions, particularly since the newly emerging youthful masculine model was perceived as 'often synonymous with embezzlement, materialism, and cynicism' (Štulhofer and Sandfort, 2005: 4).

Even though women were at first disproportionally affected by the new inequalities (being first to be dismissed from employment, for instance), they seem to have been less affected psychologically. This difference has largely been attributed to greater habitual adaptability – in tandem with well-practised traditional gender roles – as well as to a growing range of opportunities for advancement (True, 2003: 74). While this greater adaptability of women to rapidly changing socioeconomic conditions has resulted in a relatively higher rate of professional success (compared to men), it has also made visible gendered fictions of individual identity and efficacy. As the proportion of women functioning as *primary* breadwinners increased, so too did rates of domestic violence.

The process of incorporation into the EU, which accelerated in the late 1990s, mitigated some aspects of neoliberalization in the interest of long-term pan-European sustainability. Yet, by requiring structural adjustment prior to the accession of CEE countries to full membership, it predominantly extended and solidified the established neoliberal practices in the region, practices that benefited primarily Western markets and investors rather than the local people or the economic structures of their livelihoods. The requirements for CEE markets to open to foreign investment and privatization – while Western markets have remained heavily regulated and insulated – have made a paradoxical centrepiece to the conditions for accession. As Grabbe observed in 2003, on the eve of the first three accessions of Hungary, Poland and the Czech Republic to the EU:

> There are double standards, and some of the rules are more equal than others. Thus there has to be free movement of goods, services and capital in central and eastern Europe, but the EU is not going to allow free movement of people from east to west immediately (even though free movement of workers is one of the four freedoms

> of the Single Market). State aid has to be limited according to EU rules, without the laxity that has been allowed to German *Länder* or declining industries throughout the EU. National subsidies to agriculture must be reduced, but east European farmers are not guaranteed access to the major Common Agricultural Policy funds. (Grabbe, 2003: 79)

Furthermore, trade agreements conducted between the Western and Eastern European regions were marked by a distinct favour for Western markets, maintaining strict import quotas on products crucial to CEE exports, invoking let-out clauses, imposing bans on CEE agricultural products (on the grounds of divergent sanitary rules) and steel (threatening to invoke 'dumping' clauses and outlining protections in trade agreements), and objected to low labour costs in CEE despite the fact that lowering wages had been one of the EU's earlier neoliberal recommendations (Okey, 2004: 119). In many cases, the EU upheld complaints of Western companies that desired 'to maintain a previous monopoly position' (ibid.: 182). Despite rosy rhetoric of a united Europe, the accession negotiations revealed that the existing members were steering the process to include new countries as, at best, 'largely junior partners and facilitators of the running of their own economies, as adjuncts to a global hegemony of the pre-1989 west' (ibid.: 174).[3]

Not surprisingly, as the positive outcomes of shock therapy failed to materialize, the mid- and late-1990s began registering increasing popular levels of Euroscepticism throughout CEE (cf. Grabbe, 2003; 82, 88; Okey, 2004: 120, 185ff; Wedel, 1998). Reflecting the decreasing trust of their constituents in the EU accession process, CEE politicians and political parties began centring their platforms on variations of resistance to the accession requirements, usually within the pragmatic general confines of 'yes, but' (Neumayer, 2008: 135). While attempting to negotiate for a continued positive EU accession process, governments also deliberately stalled remaining privatization projects, retrenching into more protectionist positions of key industries and social services. Well-trained by long years of negotiating the dominant Soviet regime, which did not operate that differently from the remote-control economic dominance Western advisers were trying to impose, CEE politicians returned to the practice of *ochkoviteratelstvo* wherein official rhetoric, accompanied by necessary public policies, designed to please international powers did not correspond with local sociopolitical developments (Wedel, 1998: 73).

The material outcomes of the 1990s neoliberal shock therapy applied in CEE underscored not only the fallacious promise of neoliberalism to improve general living conditions rooted in the presupposition 'that free markets and free people are part of an inseparable project' (Klein, 2007: 183), but also the jarring ideological incompatibility between neoliberal practice and functioning democracy.By relegating material well-being to the responsibility of the individual, neoliberalism has tended to disempower the core function of democracy by routinely assigning inequalities 'to "private" life, which is understood as "natural", and bracketed away from consideration in the "public" life of the state' (Duggan, 2003: 5). The relegation of the individual into the private sphere has further dismantled public spaces necessary for collective democratic political action where, as Henry Giroux points out, 'individuals can be educated as political agents equipped with the skills, capacities, and knowledge they need to perform as autonomous political agents' (Giroux, 2004: xv).[4] Contrary to the early CEE assumption that the rhetoric of democratic freedom embedded in neoliberal theory and practice would ensure that the shock therapy would accelerate the transformation process of postcommunist societies in flux into Western liberal democracies, the neoliberal practices created further distance between the affluent West and the postcommunist countries. While neoliberalism seemed to be synonymous with Western existence, the shock therapy unveiled it to be an exploitive strategy of the West to benefit from the weakening democracies and economies of their eastern neighbours. Put simply, the metaphor changed only slightly: from being synonymous with the West, neoliberalism became a synonym for Western exploitation.

Acutely constrained democracy, in the CEE and elsewhere where neoliberal restructuring has been implemented,[5] has served as one of the most noticeable signs of the neo-imperial characteristics of neoliberalism, as scholars of both its theory and practice have increasingly pointed out. In the aftermath of the West's official relinquishment of its colonies, neoliberalism has come to replace (and frequently expand) the frontiers of predominantly Western economic and political exploitation. Rather than charting a new course of development, neoliberalism has merely approximated colonialism in far more elegant new clothes that retain the strings to Western-consolidated power and resources. Rather than openly extracting resources from dependent territories, for instance, the economic elites of the established West have worked through neoliberal financial institutions such as the IMF and the World Bank to charge interest on loans provided to developing countries nominally to counter balance economic hardships resulting from political or economic crises or natural disasters. Since many of these hardships developed as consequences of Western colonialism and/or subsequent Western political involvement in the regions, the loans, their exploitive neoliberal clauses and their unpredictably changing rates of interest furthered an exponential impoverishment of the 'assisted' countries while enriching the lenders. It has not been lost on observers of this exchange that this process closely parallels the 'old imperial practice' of an 'extraction of tribute' (Harvey, 2007: 74).[6] In a parallel déjà vu, instead of joining a brotherhood of nations committed to democratic structures of collective self-governance, CEE countries found themselves in a paradoxical position of weakly negotiating their voluntary submission to yet another round of imperial exploitation, conducted in the traditional names of modernization, progress and freedom.

In this context, the rhetoric of individual and human rights, such as the rights of homosexuals evoked in the early *Sonnet* anecdote in the Introduction, has become inherently suspect, particularly since it has been appropriated by standard neoliberal rhetoric. While undeniably crucial to any genuine democracy, human rights – and their rhetoric – have been repeatedly co-opted by international and local powers seeking to dominate each other. In fact, as Klein attests, '[n]o sooner had the document [the UN Declaration of Human Rights] been written than it became a partisan battering ram, used by both sides of the Cold War to accuse the other of being the next Hitler' (2007: 118). More recently, the rhetoric of human rights has been evoked internationally as 'swords of empire' to justify profoundly antihumanitarian enterprises, like waging war without a UN mandate, aimed to protect the 'interests' of the invaders in strategic geopolitical areas (Harvey, 2007: 178–9). Concurrently, as periodic public outbreaks of classified information reveal, for instance, the inhumane treatment of 'enemy combatants', it becomes obvious that Western perpetrators, despite the official rhetoric of freedom and human rights, do not use the same measuring stick to their own adherence to the human rights conventions. Within the EU, as Alexander J. Motyl has observed, similar lapses occur:

> European leaders might be willing to disregard democratic procedures if they produced results that were deemed incompatible with what the vision of Europe ostensibly represents ... [there seems to be] a double standard of moral severity vis-à-vis small states with xenophobic problems and moral lassitude vis-à-vis big states – such as France, Germany, and most recently Silvio Berlusconi's Italy – with far more vociferous and violent homegrown racists. (Motyl, 2003: 18)

In the face of obvious failures to abide by the articles of the human rights conventions in the 'core' EU member states, it may be difficult to take seriously unforgiving requirements for altering policies that are to control the observance of human rights in candidate EU countries. Moreover, the hard-line nature of such requirements, which asks for successful cultural adherence rather than a meaningful commencement of a process, seems hypocritical and prone to subversion.[7]

Furthermore, the EU's insistence on the 'human rights' of minorities, coupled with a systemic disempowerment of the citizenry, has coincided with a far more intrusive restructuring of society to suit newly rising neoliberal economic frameworks. As Jürgen Habermas has eloquently argued in *The Postnational Constellation*:

> the 'flexibilization' of career paths hides a deregulated labor market and heightened risk of unemployment; the 'individualization' of life projects conceals a sort of compulsory mobility that is hard to reconcile with durable personal bonds; the 'pluralization' of life forms also reflects the danger of a fragmented society and the loss of social cohesion ... [Neoliberalism promotes] the vision of the lifeworlds of individuals and small groups scattering, like discrete nomads, across global, functionally coordinated networks, rather than overlapping in the course of social integration, in larger, multidimensional political entities. (Habermas, 2001: 87–8)

Far from merely liberating citizen subjects from traditional modes of behaviour and social identities, the neoliberal market has *required* that potential employees become independent of stable networks of geographically limiting familial, community and social relationships. In this particular case of promoting the rights of a specific group, we could hardly argue that the recognition of social rights for the homosexual minority in itself in any shape or form influences the material circumstances of heteronormatively-identified Czech citizens. Yet the process of recognition in itself, parallel to similar legislation on women's emancipatory rights, is a byproduct of a larger process of attempting to mould individuals into more flexible units of the new economy. Put simply, neoliberal subjects are not so much liberated as made into pawns of a socioeconomic order increasingly controlled by international corporate concerns.

The growing resistance to exploitive neoliberal practices, while only occasionally visible in the EU-oriented public sphere, proliferates in a culture that digests and continually retranslates the convergences of the public and private spheres. In its multiple and multi-faceted rearticulation of political, economic and cultural capital, the cultural sphere is particularly apt at engaging the question of postcommunist identity, offering up possibilities of viable postcommunist existence in larger national and international contexts. Within the legacy of repeated colonization of the last four centuries, culture in CEE lands had become a central site where public resistance to imperial domination has cohered into a unified whole. Milan Kundera, a renowned émigré often deemed a mouthpiece of the subjugated nations, has argued that it is precisely the sphere of the CEE culture that serves as a value-repository of a repressed nation, a repository that continually serves to fuel and preserve the core of national identity until such a point when such identity can become a reality. In his bitter attack on Western Europe for forsaking CEE countries to Russian dominance, for instance, Kundera has argued that Central European societies were forcibly removed from their real Western home to preserve their Westernism clandestinely, in opposition to the dominant Soviet ideology (Kundera, 1984: 38). He has further articulated the difficulty for postcommunist CEE countries in articulating functional Western identities now that the carefully preserved 'kernels' of presumed Westernism do not seem to match the cultures in established Western countries that have been influenced by postmodernism and neoliberalism. Encountering these shifts, Kundera has insisted, postcommunist countries are left 'desperately trying to restore the past, the past of culture, the past of the modern era' (ibid.: 38). Since the predominant perception of the established West is one of cultural corruption, it is not surprising that CEE societies tend to reject existing Western models and, instead, attempt to articulate an alternative, frequently resistant mode of cultural identification.

Within the well-established lair of cultural resistance, Shakespeare has played a leading role, buttressed by long-standing CEE commitment, centuries of Western bardolatry, ready availability, sustained intellectual engagement, deep-pocketed economic investment, the purported flexibility of interpretation and a nominal lack of ideological charge. Sometimes seen as synonymous with 'culture', Shakespeare and his plays become, as Alan Sinfield has argued, its own 'signifying system through which ... a social order is communicated, reproduced, experienced and explored' (1994: 154–5). In the absence of a coherent political sphere, the cultural sphere – and Shakespeare as its significant part – comes to express values central to the new nations, providing a platform suitable for cultural and political identification, debate and dissent. Indeed, the ability to invoke and claim Shakespeare provides a semblance of transhistoric authority that may assist in articulating a cornerstone of personal, cultural and national models that anchor values loosened, undermined and put in doubt by the transitional process. In this context, Shakespeare provides a bridge between the precarious official, public, political sphere – always balancing between international powers that would claim CEE in their camp – and the individual citizen subject faced with the material consequence of the postcommunist, neoliberal, West-oriented process.

2
Shakespeares of the Postcommunist World

If a new Flood were to destroy humanity and a new Creator were to try to create anew the people that inhabited the earth before the catastrophe, He could take Shakespeare's plays as His guide and find there all kinds and conditions of men from kings to criminals, from maids transfigured by love to women repudiating their own father – yet in all that throng He would not find two people alike (not counting Viola and Sebastian or the Antipholuses and Dromios).

Jaromír Pečírka, 1966: 83

The genius of Shakespeare will have understood this phantomalization of property centuries ago and said it better than anyone.

Jacques Derrida, 1994: 42

I. Shakespeare the politician

That Shakespeare could be involved in the sociopolitical discourse of postcommunist countries should not come as a surprise, considering the engagements of Shakespeare's texts in nation-building (and nation-bashing) processes around the globe since the late eighteenth century. Numerous scholars of the cultural-materialist (and various derivatives thereof) creed have delineated constructed qualities of the entity called 'Shakespeare'. Far from merely denoting a playwright or, in the wider meaning, a set of texts written by a playwright, this entity has functioned as a site of ideological contest for 'owning' Shakespeare's presumed cache of essential humanity and truth to further particular sociopolitical agendas. The danger of positioning Shakespeare as what Jonathan Bate called the 'Dead White European Male in chief' (1994: 124) and Terrence Hawkes recently labelled the 'industrial-strength insect-repellent' (2002: 142) lies in the use of Shakespeare as an extendable arm of exploitive Western practices, ranging from the classical colonial regimes of the eighteenth and nineteenth centuries to current neoliberal practices.

Shortly after the fall of communism in CEE, Hawkes provided an exemplary – if deliberately inflammatory – summary of the underlying assumptions traditionally made about the truth that Shakespeare presumably embodies:

> 1. That human nature is permanent, one and indivisible, regardless of place, race, creed and culture. In the end, under the skin, we are all the same and it is to this sameness that Shakespeare speaks.
> 2. That passage of time, history, makes no difference to this.
> 3. That, construed aright, and analysed with sufficient ingenuity, application, vigour and flexibility, Shakespeare's plays are able to address all people at all times, and everywhere.
> 4. That to deny any of the above is to reveal serious deficiencies in one's humanity, such as characterize the ravings of the perverse, the envious and the politically and socially deviant. (Hawkes, 1996: 10)

This construction of Shakespeare's universality serves as a founding tool for a multiplicity of political purposes, where ambiguous and multivalent reality can conveniently be collapsed into a simplified model. The insistence on universal sameness has often served to erase meaningful categories of difference and their implied potential to challenge the hegemonic status quo, while the requisite correct interpretation (#3) has provided institutional scholars with the power to define normative parameters of human existence. Even arguments relating to Shakespeare's complexity betray a desire for a singular, graspable model of an idealized view of reality that could be mastered, learned and translated into relevant practice. The fourth and final assumption that Hawkes describes is particularly revealing of the framework through which idealized Shakespeare stands to gain actual political power, through the division of humanity into categories of artistic and universal cognizance. Such divisions can be converted tidily into a rationale for cultural and economic domination, exclusion and 're-education'. In other words, those with the power to define Shakespeare's truth also gain the power to categorize others as deviant in a variety of 'other' categories and therefore to seize considerable political power.

Shakespeare's universality in this context provides an authoritative ideology that can be used as a basis for universal repressive hegemony. This ideology, as Jean Howard and Marion O'Connor have suggested, 'can never be "disinterested" because it functions to render "obvious" and "natural" constructions of reality which, often in oblique and highly mediated ways, serve the interests of particular races, genders and classes within the social formation' (1987: 4–5). They further argue that:

> [p]robably more than any other figure in Western culture, Shakespeare has been used to secure assumptions about texts, history, ideology, and criticism ... He functions ... as a kind of cultural Esperanto, a medium through which the differences of material existence – differences of race, gender, class, history, and culture – are supposedly canceled. He is repeatedly presented as the writer who transcends such differences to get at the abiding truths of human existence, as 'not for an age, but for all time.' Claims about Shakespeare as the bearer of universal truth serve an oppressive function when they render illegitimate readings produced outside the dominant ideologies which secure a society's understanding of what the true is. (Ibid.: 4–5)

Shakespeare's assumed stakes in truth thus not only dislocate claims made by unworthy interpreters, but also undermine cultures that function differently from the established West. Shakespeare's enforced universality has been harnessed simultaneously to eradicate existing differences between human contexts and to insist on Western cultural norms as the universal template. More specifically, Howard and O'Connor suggest that Shakespeare's universality has been used to simplify complexities of human existence to a limiting perception of the world driven by a 'dominant ideology'. Traditionally speaking, this ideology has supported the claims and worldviews of the European, white, male-centred, affluent, Christian, heterosexual and colonial powers that reserved the right to exclude from 'universal humanity' all those who fail to comply.

The political dimensions of handling Shakespeare are often masked by a wilful denial of ideological links between art/culture and politics. Describing the ingrained nature of this separation that seems to make it 'common-sense', Hawkes delineates the invested production of oppositional discourses of art and politics that are to serve as the founding basis for nominally oppositional areas of organized human existence. In his usual cheeky manner, he posits that any suggestion of common ground between art and politics 'is almost to sanction some illicit act of transgression in which a grubby "Politics" may be "dragged" across a threshold to sully the otherwise sacrosanct shrine of Art. Worse, reversing the process, Shakespeare himself might even be "dragged into politics"' (Hawkes, 1992: 43). More kindly, Jonathan Dollimore classifies such divisive approaches to Shakespeare as 'idealist' within the spectrum of possible claims about the relationship between literature and politics. This idealist view seeks to protect art as a haven from the political: 'the precious thing about art is that it transcends the sordid world of politics, reaching toward a realm of enduring truth and value which politics can at best ignore, at worst contaminate and betray' (1998: 260).

The fallacious separation of Shakespeare from politics is well-illustrated by scholars researching Shakespeare productions in colonial and postcolonial societies, who repeatedly point to the practice of using Shakespeare as a tool of ideological oppression working in concert with the more forceful military, economic and political repressive regimes. In this context, Shakespeare has been evoked as an imposed paragon of Western civilization, unmatchable by the cultural achievement of the colonized populations, even as those colonized explored the possibility of subversive use of these tools for their own ends in resisting the imposed hierarchical social framework. On the one hand, Shakespeare has been used by colonial powers to exclude and dominate colonial subjects. On the other hand, those who were colonized frequently sought to prove mastery of Shakespeare as a bid for inclusion and equality within the dominant sociopolitical and economic discourse. For instance, Ania Loomba and Martin Orkin take for granted that Anglo-American literary scholarship of the last two centuries offered a Shakespeare who celebrated the superiority of the 'civilized races' and that:

> colonial educationists and administrators used this Shakespeare to reinforce cultural and racial hierarchies. Shakespeare was made to perform such ideological work both by interpreting his plays in highly conservative ways ... and by constructing him as one of the best, if not 'the best,' writers in the whole world. He became, during the colonial period, the quintessence of Englishness and a measure of humanity itself. Thus the meanings of Shakespeare's plays were both derived from and used to establish colonial authority. (1998: 1)

In such a setting, they add, 'reinterpreting Shakespeare's plays became, at least for some critics, part of the business of reinterpreting and changing our own world' (ibid.: 3). Simultaneously, as Thomas Cartelli attests, dominated societies 'refashion' Shakespeare 'outside the national boundaries of the British colonial culture and society in the image' of its own both to 'establish their independence from imperial influence' and 'to identify, define, and assert their own national values or priorities' (1999: 2). In many cultures, therefore, translations, interpretations and productions of Shakespeare have served as proof of 'cultural credibility' and also the credibility of language that were to prove that the development of the culture in question was on a par with its English models. As David Quince has noted, Shakespeare has been perceived as 'a litmus test of civilized culture' (2000: 80).

Shakespeare's litmus characteristics are not limited to explicit (post)colonial contexts, but function equally informatively in the wider cultural competition for inclusion in the economically and politically powerful Western bloc. An intriguing – though not isolated – example is that of Japan, which has struggled to overcome its Second World War defeat to become one of the modern global (predominantly Western) economic and political heavyweights of the later twentieth century. Speaking on behalf of the Japanese in the introduction to the published proceedings of a seminar at the 1991 World Shakespeare Congress in Tokyo, Minoru Fujita denies his culture's commitment to Shakespeare, only to embrace immediately and wholeheartedly the cultural capital that proficiency in Shakespeare offers. On the one hand, he claims that '[n]ever once ... did the Japanese people experience such a cultural upsurging induced by Shakespeare as was found in eighteenth century Germany, nor did they ever so unanimously find his work so obsessing to their souls as to urge them to claim that Shakespeare belonged to them' (1996: 3). On the other hand, he reminds his Western readers that Tokyo is a site of a Globe replica and, pointing to the rising number of Shakespeare productions in the last two decades, he provides evidence of appropriations just as multi-layered and politically complicated as in other non-Western cultures with ambivalent relations to the Western 'core'. The disclaimer of Japanese interest in Shakespeare and the objectivity implied therein paradoxically serves as evidence of the unique privilege of the Japanese access to 'true' Shakespeare through the traditional venues of Japanese theatre:

> Every possible aspect of Shakespeare's drama ... has been thoroughly and vigorously studied in Japan, and no Shakespearean scholar in Japan will deny the overwhelming superiority of Shakespeare's drama as literature ... Paradoxically, Shakespearean scholars in Japan who are truly aware of the excellence of Kabuki and other theatrical traditions of their country are ... gifted with a particular deep insight into the timeless value of Shakespeare's drama as theatre art, and therefore have become more interested in the unalloyed, definitive representation of genuine human passion or true vital force to which each of Shakespeare's plays is originally meant to give exhaustive expression. (Ibid.: 7–8)

Fujita's rhetoric exemplifies a larger competitive discourse of nationalistic Shakespeare appropriation where the capital of 'true' Shakespeare parallels competitive cultural achievement, which subsequently lays claim to political and economic capital on a global scale.

This competition of ideological and material capitals via Shakespeare trickles down, as Andrew Gerstle has suggested, to a culture's perceptions of the validity of its claims to globally-exerted economic and political power. Hypothesizing about the correlation of the degree of Shakespeare bardolatry and national politics in non-Western contexts, Gerstle posits that the ability to claim that one's cultural products contain 'universal elements' and to challenge the standing

Euro-American universals comes in a 'direct correlation' between a culture's 'self-confidence' and its 'socio-economic might' (1996: 65–8). This 'cultural self-confidence' can be gleaned by surveying the fervour of Shakespeare worship in a given culture. In the case of Japan, the relatively recent Japanese interest in Western culture, particularly Shakespeare, seems to be a natural consequence of Japan's defeat in the Second World War, after which the nation sought to re-establish itself culturally as 'an honorary member of the Western civilization' through emulation – rather than critique – of 'the art of a powerful and civilized society' (ibid.: 64–5). Consequently, Gerstle argues, 'the extent of [Shakespeare's] world-wide popularity has had far more to do with the economic, political and military might of both the British and American empires than with the inherent genius of his plays' (ibid.: 69).

II. CEE's political Shakespeare

Because Eastern Europe had not formally fallen under British rule, Shakespeare has not been used as an explicit tool of English-language colonial domination. Instead, his works afforded an internationally well-recognized cultural capital that seemed divorced from the immediate sociopolitical circumstances and therefore divorced from local politics altogether. As such, this repository of apparently free-floating transcendental culture could be mined for insights about 'humanity' and 'truth' that reached beyond the temporal political moment and, if used with grace, could be called on by both colonial systems and colonized societies to compete for cultural superiority. During communism, this competition intensified; while Shakespeare remained nominally non-political outside the struggle between native cultures and the Soviet Union, the Cold War context underscored Shakespeare's subversive belonging to the side of the Iron Curtain opposite to the communist regimes, despite repeated attempts by communist rhetoric to claim Shakespeare as an early communist mired in repressive (Elizabethan) early capitalism.

Much of the actual cachet of Shakespeare's cultural capital under communism was enabled by the reality that Shakespeare was difficult to censor. As an author of undeniable ideological power in late nineteeth- and twentieth-century Britain, where his admirers Karl Marx and Friedrich Engels produced most of their communist tracts, he posed a problem for communist ideologues in charge of defining an artistic canon that would suitably express the tenets of socialist realism. Eventually, they agreed on an elegant appropriation of Shakespeare to the Marxist canon via a careful interpretation of his texts that centred on early versions of the 'socialist hero' and proto-working-class consciousness. At the 1934 Congress that established socialist realism as the dominant prism for all communist-approved art, Shakespeare was so firmly established at the heart of the communist canon that his giant portrait, together with an equally imposing portrait of Maxim Gorkii, hung over the assembled congress delegates; writers and dramatists contributing speeches 'gave their oath of allegiance' to the two portrayed artists, vowing to abide by the new socialist realism guidelines (Ostrovsky, 2006: 56).[1]

On the heels of such official approval, buttressed by Stalin's own admiration of the playwright, Shakespeare was performed throughout the 'second world'. Though his productions were closely watched for subversion and topical references were uniformly forbidden, his plays were officially approved as appropriate material. As Zdeněk Stříbrný argues, '[e]ven the dyed-in-the wool apparatchiks did not dare to attack him openly, although they found it personally offensive to hear from Hamlet that something was rotten in the state of Denmark' (2000: 97). Inevitably, Shakespeare became a site of subtle (and sometimes not so subtle) ideological war, wherein official interpretations sought to use Shakespeare and his work as ideological tools against Western capitalists, while many of the performers and scholars attempted to present a Shakespeare that undermined official ideology and reconnected with 'the West', a shorthand symbol of a 'free' world.

In addition to pressing questions of national independence and individual freedom, cultural competition remained central to much Shakespearean scholarship published under communism. Whether produced by friends or foes of communist regimes, the reams of essays and monographs exemplary of this era fall into two broad categories. On the one hand, evaluative reviews of Shakespeare productions and translations seek to ascertain whether they succeed or fail in communicating the playwright's primacy of artistic transcendental achievement. On the other hand, erudite readings of Shakespeare's plays underscore specific utterances, plots and themes particularly relevant to the tension between the individual and the collective, the self and the state, ethical responsibility and state ideology. *En masse,* the scholarship becomes a vigorous blur of seemingly repetitive arguments, all of which passionately seek to deal a painful blow to the opposing side's ideology while enlisting the support of internal and international like-minded intellectuals. As current scholarly research on Shakespeare produced under communism makes clear, the evaluative arguments and erudite close readings reached far beyond mere intellectual play, providing an outlet for otherwise silenced intellectual figures eager to sneak a dissident voice into the heavily censored public sphere. Shakespeare productions and scholarship, read closely by the authorities and intellectually-voracious audiences, fuelled the fortunes of all participants, earning them an unpredictable variety of rewards ranging from party promotion to international intellectual acclaim, enforced manual labour, Siberian exile and execution.[2]

On the heels of the predominantly peaceful revolutions of 1989, Shakespeare re-emerged at the forefront of CEE cultural production, victoriously proclaiming the timelessness and universality that would buttress the presumably impending CEE 'return to Europe'. A collection of essays published in the early 1990s, appropriately entitled *Shakespeare in the New Europe*, provides an instructive example of the projection of postcommunist sociopolitical developments into the production and scholarship of Shakespeare. Reflecting the hopeful moment of postcommunist elation at the time of publication, the collection includes essays by Shakespeare scholars from both sides of the recently-fallen Iron Curtain brought together around an agreement that 'Shakespeare was "not of an age, but for all time"' (Hattaway *et al.*, 1994: 19). Many of the volume's CEE contributors demonstrate the turn to Shakespeare to seek the 'truth' so central to postcommunist stability. Mark Sokolyansky, for instance, evokes Shakespeare as the moral cornerstone model for postcommunist society, counterbalancing the chaos resulting from the rapid dismantling of the communist regime: 'Shakespearean concept of society and the processes of its development can be a guide for the new generation of east European statesmen, or – even more – for conscientious citizens' (Sokolyanski, 1994: 224).[3] Similarly, in her discussion of the political transitions in postcommunist Romania, Odette-Irenne Blumenfeld reports that Shakespeare appeared as the first author on the postcommunist stage, as 'his texts have always appealed to a society in a process of change' (1994: 232).

Even in societies where the postcommunist era brought social unrest and even armed conflict, such as the former Yugoslavia, Shakespeare remained centre stage and interpretations of his texts called upon his singular universality to provide a guiding light in a time of profound difficulty:

> the presence of Shakespeare after 1989 in the countries of the former eastern Europe reinforces the appeal of the Shakespearean texts to a society in transition, in search of an identity. In their confusion and chaos, people turn to Shakespeare as a permanence, capable of bringing home to them allusions to the complexity of today's background, thus placing it in a historical perspective, offering reassurance by underlining what is eternally valid in human experience. (Hattaway *et al.*, 1994: 17–18)

Despite the clear parallel between political events, subsequent changes in the CEE sociopolitical climate and the publication of this collection, a defining characteristic of the CEE essays is a staunch denial of the political dimensions of postcommunist Shakespeare. Despite Shakespeare's explicit involvement in postcommunist cultural redefinition parallel to political and economic reconstruction, an involvement that many of the essays describe at length, most of the authors maintained that postcommunist Shakespearean production has lost its political edge. Speaking of postcommunist Shakespeare theatre in Bulgaria, for instance, Alexander Shurbanov and Boika Sokolova argued that '[t]he owners of the first private theaters in eastern Europe ... would rather keep the public well entertained and not provoked into thinking – at least until they have established themselves as the new masters of the said public' (Shrubanov and Sokolova, 1994: 53). The consensus seems to be that once the revolutionary dust settles, Shakespeare will be safely relegated to the realm of pure culture, no longer tainted by the polluted sphere of public politics. It is ultimately this desire to cleanse culture of politics that articulates the hope that the CEE Shakespeareans held for the upcoming Westernizing process: the submersion of political commentary in Shakespearean scholarship and production during communism had been a thing of necessity, utilized in response to restrictions on free speech in the public sphere; the hope for working democracy promised to remove this necessity, leaving Shakespeare to soar freely towards politically-untainted artistic heights. Moreover, since Shakespeare had been widely heralded as a proponent of universal humanistic principles, his works were expected to coincide with the building of an imperially-unhindered postcommunist society.

Yet, as scholars witnessing the unfolding neoliberal shock therapy in CEE have noted, a harmonious democratic development in postcommunist Europe was not (at least immediately) to be. It is telling, as much as it is ironic, that Jacques Derrida's first self-consciously and forcibly political work of scholarship, *The Specters of Marx*, that surveyed the likely global cultural developments in the aftermath of a neoconservative celebratory proclamation of the death of Marxism in 1989 is haunted – even besotted – by Shakespeare. The spectres of Marx that inevitably haunt Europe to likely re-emerge in rearticulated guises as a meaningful response to neoliberalism in Derrida's work have become synonymous with the wronged ghost of Hamlet's father demanding revenge from the living.[4]

III. Czech Shakespeare and the history of colonial resistance

The remainder of this chapter will consider the specific trajectory of Czech Shakespearean appropriation harnessed in moments of significant political upheaval resulting from periodic waves of Czech resistance to imperialism. Namely, I will survey Shakespeare's reappearance in three transitional moments of Czech sociopolitical development, ideologically intertwined with each other in constant reference to transitions that have already unfolded and to transitions that may come. My entry point into this transitional Shakespearean analysis is an informative volume, *The Charles University on Shakespeare*, published in 1966 as the totalitarian Stalinist influence of Moscow weakened and the Czechs began imagining the possibility of alternative social organization before Moscow reasserted its dominance in the summer of 1968. I begin with a detailed consideration of the ways in which the volume's contributors harnessed Shakespeare to present their political positions in the relatively loosened political climate. I follow with an analysis of the contributors' frequent citations of the 1864 Shakespeare tercentenary that coincided with the height of the Czech National Revival, a movement which sought to win Czech autonomy from the Austro-Hungarian Empire. Finally, I turn towards the transitional moment of the postcommunist period. In considering each transitional turn to Shakespeare, I examine the vital cultural importance of intellectuals participating in the transitional process by a firm insertion of the cultural and artistic sphere into the sphere of public politics. Within this inclusion of 'culture' in the public sphere, I chart a persistent practice of claiming Shakespeare's 'truth' as inherent to the Czech transitional course. Shakespeare's universality is moreover evoked to buttress his status as a non-political entity that impartially enters the public sphere to champion Czech rights to national independence and cultural-self determination against the Austro-Hungarian Empire in the 1860s, the USSR in the 1960s and the EU in the 1990s. Finally, I attempt to show how, in each transitional period, the issue of individual citizenship and national identity, mediated via Shakespeare, plays a central role in the rhetoric of political and economic restructuring.

The Charles University on Shakespeare, a slim, elegantly-bound English-language book, with a prominent 'WS' and the university crest on the cover, presents the proceedings of a 1964 Shakespeare conference hosted by the university to commemorate the quatercentenary of the playwright's birth. Its contributors – without exception current or past professors at the institution – represented a wide range of disciplines, spanning from the predictable faculties of drama, English and American studies to history, classical and modern philology, art history, and Czech language and literature. The conference and subsequent publication overtly aimed, as the editor notes in the introduction, 'to show the interest in Shakespearian studies of the different humanities departments of Charles University and of all present generations of University teachers' (Stříbrný (ed.), 1966: 9). The less readily articulated – but nevertheless palpable – effect was a multi-disciplinary manifesto outlining an intellectual shift in the conceptual framework of categorizing lived reality, a shift that sought to break away from hegemonic communist ideology via Shakespearean cultural capital evoked simultaneously as the scalpel of political criticism and an antidote to watchful communist censorship. As such, the volume provides an excellent example of transitional political nation-building negotiated not through explicit traditional political channels, but via the medium of culture articulated through Shakespeare. The collective contributions of the leading university intellectuals, perceived in the Czech context as integral keepers of the core of the imperially-besieged culture, further exemplify the active progress of transition from the sphere of improvisational practice (seemingly ad hoc subversive theatrical practices that sought to clandestinely disperse anti-totalitarian images in Shakespeare productions) to the sphere of institutionalized publicly-shared theory (published and professionally-vetted reviews and scholarship) that contributed to a surfacing reformist nation-building process.

The volume appeared at a time of feverish transition enabled by a period of respite in communist vigilance of the 1960s. Though not ready to abandon communism altogether, the Czech regime explored forms of socialism alternative to totalitarian Stalinism eventually to pursue a new form of 'socialism with a human face'. Eagerly supported by a great majority of the population and the government, keen to create maximum distance between the Czech political system and the recently revealed terror of Stalin's rule in the USSR, this effort culminated in the so-called Prague Spring of 1968 when widespread opposition to unethical communist practices led to substantial changes in the highest levels of the Czech government. Heeding public pressure, President Antonín Novotný resigned his post in favour of the popular, promptly-elected Ludvík Svoboda. Svoboda brought with him a new, multi-party government that, though predominantly leftist, was no longer exclusively communist. The new socialism with a human face progressed with much popular support until a Soviet military invasion on 21 August forcibly returned the system to Moscow-controlled tracts. The subsequent 'normalization' of the Brezhnev/Husák era propagated a retrenched version of the earlier system, now dubbed 'real socialism'. This forced regression, overseen by Soviet troops stationed throughout the country, brought on a profound mass disillusionment with the official political system, a disillusionment that eventually led to popular protest in late 1980s. In 1966, however, when *The Charles University on Shakespeare* was published, the hope for real change – an early version of perestroika fuelled by the hope of a return of democratic free elections – was beginning to take hold.

In such a political climate, *The Charles University on Shakespeare*'s collective intellectual voice could be published, though still within the cautious constraints of established cultural discourse. Thus, instead of outlining explicit political directions, the contributors analyse presumably admirable social arrangements in Shakespeare's texts, the nuances of his presumed views on the Elizabethan social hierarchy and his 'likely' views of communism. Those who were more daring present copious outlines of the Czech Shakespearean tradition to offer corrective suggestions to established interpretive models championed by the communist regime. In seeking to negotiate successfully the hegemonic ideology of an oppressive imperial system, bardolatry – a passionate worship of Shakespeare as the pinnacle of humanity and artistic expression – was requisite. But rather than being a blind and unquestioning admiration for the playwright, it served as a keenly political tool of subversion that deliberately claimed Shakespeare's superiority as a repository of alternatives to the present *modus operandi*. Particularly in a context where Shakespeare's cultural capital was evoked for independence-minded self-determinatory nation-building purposes to counteract dominant and derogatory imperial ideology, claims to his transcendental universality assisted in trumping the temporal claims of the current empire.

Despite the inevitable individual focus of each contribution, the volume as a whole underscores several currents powerfully propelling the reformist discourse: an emphasis on common humanity as a basis of transnational solidarity, a resistance to singular intellectual inquiry embodied in communist ideological hegemony and an insistence on 'truth' as the only viable organizing principle of everyday reality. These three currents operate in concert to argue for the necessity of articulating alternative modes of existence to the forcefully optimistic, ideologically-driven socialist ideals championed by the communist system. The quest for 'truth', in particular, common not only to the contributors to the volume but also to the general genre of dissident writing, betrays a fervent longing for breaking down the doublespeak of the state that confines the parameters of everyday experience of each citizen subject. In its place, the writers argue for a mode of existence aligned with an assumed internal individual identity, a mode of existence that is widely defined by freedom of expression and from surveillance. The field of Shakespeare studies and performance provided an ideal space for piloting such resistant discourse. Shakespeare's well-cultivated presumed transcendental value has assumed that the much-needed 'truth', which is to lead reformist models of existence, lay within the texts themselves. The quest for existential liberation was thus projected into the method of textual interpretation pitting the communist-dictated socialist realism, established at the 1934 Congress of Soviet Writers as the sole appropriate mode of producing and interpreting art for the 'education' of citizen subjects, against any other interpretive method that would produce alternative readings of the texts.[5] Yet, crucially, though this model – socialist realism against anything else – would suggest a quest for democratic multiplicity, the process was far more interested in an eventual articulation of a coherent (and therefore relatively singular) alternative. A stated, well-maintained cultural commitment to Shakespeare would further ensure that resistant Shakespearean scholarship and theatre work would not remain isolated within the sphere of 'culture', but would be integrated into the wider cultural discourse of budding dissent.

A short selection from the contribution of Jaromír Pečírka, a professor of art history at the university, to *The Charles University on Shakespeare* volume provides an excellent example of a bardolatrous interpretive moment that suggests a corrective to the dominant communist system that would reflect a more 'true' – in this case nominally universal – mode of social organization:

> If a new Flood were to destroy humanity and a new Creator were to create anew the people that inhabited the earth before the catastrophe, He could take Shakespeare's plays as His guide and find there all kinds and conditions of men from kings to criminals, from maids transfigured by love to women repudiating their own father – yet in all that throng He would not find two people alike (not counting Viola and Sebastian or the Antipholuses and Dromios). (Pečírka, 1966: 83)

Pečírka's evocation of Shakespeare as the paradigm of the varied template of humankind undeniably reveals a world diametrically different from the everyday reality imposed by the communist totalitarian regime. The mere invocation of a biblical 'Creator', banned by communist ideology, here transcends all temporal and secular authorities, including the one that denies him. Humanity here is not defined by a homogenizing communist ideology but by an assumedly-universal Shakespearean model. Rather than representing a horizontal scale of difference championed by egalitarian communist rhetoric, the world according to Shakespeare betrays a hierarchical model reminiscent of the great chain of being: males rank on a vertical scale of the social order, from those at the top ('kings') to those relegated to the bottom ('criminals'). Similarly, females do not appear as independent agents but are signified by their relationship to men: virginal maids 'transfigured' to a presumably ideal state of being by the love for their men are opposed to sexually nondescript 'women', incidentally in Pečírka's sentence structure parallel to criminals, who defy their primary patriarchal source of authority, their father. Such an ideal universe flaunts central pillars of communist ideology, wherein 'kings' have become enemies of the working class if not criminals, and women have become independently-defined subjects, measured by their commitment to the proletariat and its socialist goals. Perhaps ironically, Pečírka's 'perfect' world order that is to reflect 'true' everyday experience, which he tentatively substitutes for his everyday reality, does not entirely strip frameworks of ideological reference championed by dissidents to communist ideology. Rather, this articulation of a Shakespearean alternative to communist hegemony staves off the vestiges of the communist state – here symbolized by the state's insistence on the nominal autonomy of equal citizen subjects – to impose alternative frameworks of human categorization, even if that means a return to traditional vertical forms of patriarchal power throughout the society.

Even as this interpretive moment – within the context of the entire volume – undeniably signals an unprecedented (if still tentative) freedom of expression uncensored by the state authorities, it simultaneously exemplifies the easy conflation of anti-communism with 'truth' and transcendentalism. While desperately seeking to escape the totalitarian control of ossified communist ideology, continuously championed by the system despite the glaring inconsistencies between its premise/promise and everyday reality, many dissidents find 'truth' in the act of mere *opposition*, that is, the binary opposite of the position staunchly held by the communist system. While the overall thrust of the transitional moment of the 1960s was predominantly reformist – looking to rearticulate the present system into *socialism* with a 'human face' or a leftist system capable of delivering on the initial promises of the now seriously disjointed and corrupt Soviet form of communism – this framework of binary opposition frequently led to an unquestioning affinity with the nominally 'free' world, vaguely but undeniably located in the capitalist and nothing if not ideologically driven West.

IV. When in doubt, celebrate Shakespeare

Jan Kott's *Shakespeare Our Contemporary* (1964) became a powerful weapon of the 1960s Czech reformist intellectuals. Banned but widely read by CEE intellectuals, his focus on the darker aspects of Shakespeare's plays and parallel questioning of Shakespeare's most cherished hero-like characters undermined the required optimism of the approach of socialist realism. When Kott's essays were published in Czechoslovakia in early 1964, the political system was already under popular scrutiny and the volume greatly contributed to the process of questioning the current ideological hegemony. It is telling that in that same year, Kott attended a translation conference held at the Prague theatre Na Zábradlí, where he presented his views both on Shakespeare and on relevant possibilities of reforming the existing communist regime to chief Czech intellectuals from the university and theatre worlds.

Kott's internationally recognized scholarship, fuelled by a bitter disappointment with the violent corruption of Stalinist communism admitted publicly in 1956, refused to recognize Shakespeare's genius in his works' optimistic outlook but maintained, instead, that his transcendental value lay in his uncanny comprehension of the Grand Mechanism of history that 'has no meaning and stands still, or constantly repeats its cruel cycle' made mostly of 'gigantic slaughter' (Kott, 1967: 32, 40). In the place of a celebration of a certain future victory of the working class over exploitive, proto-capitalist hierarchical systems, Kott celebrated Shakespeare's ability to reflect accurately the general madness of humanity prone to violence and meaningless pleasures, divided into two simple categories: 'those who kill and those who are killed' (ibid.: 77). As the Shakespeare volume testifies, however, the chief attraction of Kott's work to Czech intellectuals was not its fatalistic outlook but the staunchly alternative Marxist – explicitly *not* socialist-realist – point of view which provided a useful alternative inroad in approaching the cultural capital of Shakespeare to produce alternative laudatory articulations of his genius and, subsequently, alternative models for the categorization of reality. Kott further reinforced this crucial multiplicity through the fatalism of his view on Shakespeare interpretation; claiming that all interpretations will inevitably 'butcher Shakespeare' since 'Shakespeare is truer than life', Kott inadvertently dismantled the supremacy of the communist single-minded ideology-driven compulsory mode of interpretation. His momentous contribution climaxed with the removal of Shakespeare from the historical perspective of a lofty forefather to place him, via his underscored universality and transcendence, as 'our contemporary'. Thus primed, Shakespeare became poised to yield genius advice to the current state of affairs, a reality not lost on the contributors to the university volume who took an opportunity to tap Shakespeare for insights relevant to the particular Czech political context.

The spirit of Kott's momentous work permeates the majority of the contributions to *The Charles University on Shakespeare*, which subsequently generate diversified frameworks anticipating the hatching transitional process. Bohumil Trnka, for instance, evokes Kott by insisting on a 'constant need to reassess' Shakespeare within a 'realistic understanding of his work free of all idolatry' (1966: 55, 56) to access Shakespeare's 'ethical realism' (ibid.: 62). Such a matrix of humanistic ethics (implicitly opposed to corrupt Stalinism), far more 'true to life' than ideologically-driven communist interpretations of Shakespeare, could serve as a basis for a reformed socialist society and play a significant role 'in the moral education of our young people, where it is so much needed' (ibid.: 63). Evoking Kott's 'contemporary' Shakespeare, in turn, Jan Kopecký looks to weaken the Soviet hold on the Czech culture via Shakespeare's implicit allegiance to the West. Indeed, Kopecký suggests, Russians seem to have missed the greatness of Shakespeare altogether, since 'Shakespeare found no permanent place in the repertoire of the Moscow Art Theatre, [and] individual productions for the most part were unsatisfactory' (1966: 108). In Kopecký's view, Stanislavsky's *Othello* merely presents evidence of 'persevering but vain effort by a great theatre personality to master Shakespeare, and a clear illustration of the conflict between conceptions of theatre [Russian and Renaissance English], one of which can only be carried out by negation of the other' (ibid.: 108–9). It is Stanislavsky's Russian background – here made inherently incompatible with transcendental universality of Shakespeare's Westernism – that makes him ineligible to master the Master. Instead of following Soviet artistic models that have failed the test of Shakespeare, then, the Czech culture is in dire need of new creative templates from the 'wider [read Western] world' to benefit 'the playwright and the public' (ibid.: 111–12). This politically charged, cultural binary of Soviet and Western Shakespeares is similarly evoked by Vladimír Štěpánek, who proposes that 'a new and more objective approach' can be gained via a realignment of the Czech culture with the national literature of other nations, 'particularly the west' (Štěpánek, 1966: 127). Only such new objectivity will allow 'Czech writers to achieve original, individual, truly national forms which would at the same time stand comparison with the literature of highly cultivated peoples' (ibid.: 134).

Perhaps the most radical contribution to the volume came from the internationally-established Shakespeare scholar, Zdeněk Stříbrný. While he lauded the 'great' Czech Shakespearean tradition, he also called on his audience to use Shakespeare as a model of resistance to an intellectually oppressive regime: 'Can we stand up to [Shakespeare's] free and fearless poetry, which like Hamlet speaks daggers; or shall we flee cowardly before it, like King Claudius?' Evoking Hamlet's painful dilemma, Stříbrný sought to awaken his listeners to the reality of everyday existence in the malfunctioning communist system, placing the burden of rectifying spreading injustice and corruption on his listeners' Hamletesque shoulders. He thus evoked the spectre of Shakespeare to provoke corrective political action. Since the central assumption, at least in this play, was that 'something is rotten in the state of Denmark' (a favourite moment of emphasis for most Czech productions during the communist years), individual recognition of such political rot called for an effort towards a collective and committed unmasking of dishonest structures of systematic behaviour.[6] Wilful ignorance that failed to challenge corrupt communist practices became a criminal offence to be avoided by following the heroic example of Shakespeare who 'does not gloss over the terrible conflicts in English history, but lays them bare as a warning' (Stříbrný, 1966: 29).

To hearten his audience for the forthcoming struggle, Stříbrný closed with a benediction in the name of 'Shakespeare's 'fair Bohemia' [and] a new wave of interpretation of Shakespeare':

> May it wash away all that is stiff and uncreative, all that blocks the way forward. May it help to take a critical and unsentimental view of all we have done ourselves so far ... Let us keep our own judgment, our own standpoint. Let us not yield those values in our own tradition and that of other countries, which are precious and

> can go on growing. And above all let us not yield the humanist values in the work of Shakespeare himself, whose end, both at the first and now, was and is, to hold as 'twere the mirror up to nature' and to tell us the whole truth about man and his life ... For today, perhaps more than ever, we need the Shakespearian synthesis of humanity, objectivity and harmony reached even in the midst of the most pitiless storms of time. (Ibid.: 37)

In this call to arms, Shakespeare emerges as the universal force substituted for the deity to which the scholar appeals, the eternal judge who holds the scales of universal justice which will measure Czech ethical and political development. Until freedom arrives, Shakespeare effectively becomes a fort, a vault which guards the precious tenets of 'humanist values' that are to provide a template for the future, revised society to follow the dark period of failed, bleak communism. As such, Shakespeare protects not only the central essence of the West, but also of 'our [Czech] tradition' that is part of the global human 'nature'.

Parallel to rousing troupes at home, Stříbrný drummed up the support of international cultural capital. Within weeks of delivering his inspiring lecture to his Charles University colleagues, he was invited to address a distinguished audience of predominantly Western Shakespeareans at the US Folger Library. A recent commemorative edition of Stříbrný's work, collected in *The Whirligig of Time: Essays on Shakespeare and Czechoslovakia* (2007), pairs the two lectures together as two sides of an identical coin. While the address to the domestic audience seeks to form a collective challenge to the communist regime via Shakespeare, the Folger lecture underscored the Czech long-standing commitment to Shakespeare to cement the nation's eligibility for acceptance as an equal to other Western powers represented in the audience. Opening with an outline of the recent Shakespeare quatercentenary in Prague, the essay firmly establishes Czechs as properly appreciative of Shakespeare's cultural capital, which they 'have come to venerate with an almost supernatural awe' (2007: 167). Here Stříbrný carefully links the fate of the Czech nation with the playwright, whose works as well as persona, Stříbrný attests, were habitually evoked on every tumultuous occasion of nation-building. Significantly, this key transcendental nation-building tool resurfaces again to assist in transcending the errors of

Stalinist communism in the 'growing recognition of political errors in the recent past, [paralleled by] Hamlet's outcry "the time is out of joint"' (ibid.: 172). Yet both sides of the Iron Curtain prove needful of thorough celebrations of Shakespeare's quatercentenary that will provide 'the humanizing touch is most needed both in the West and East' (ibid.: 175). At the forefront of this humanizing process, Stříbrný suggests, are the celebrations in 1964 Czechoslovakia, where 'Indeed, there is no end of Shakespeare' (ibid.: 174).

As the common denominator of 1960s reformist Shakespeare scholarship reveals, Shakespeare became an inherent part of hopeful discourse of political transition, a discourse aimed at transcending political demarcations of the artificially delineated 'first,' 'second' and 'third' worlds. Such discourse was to claim a common humanity with a symbolic 'free world' in the hope of securing international recognition for Czech nationhood as well as of raising political consciousness – and provoking subsequent dissident action – at home. The two 1964 lectures by Stříbrný thus exemplify the dual approach of this use of Shakespeare to argue for international support of potential separatist efforts of the Czechs from the USSR and its ideology. Simultaneously, as the domestic address demonstrates, Stříbrný and his colleagues use Shakespeare's quatercentenary to chart a new intellectual course for Shakespeare studies conceived not only as a sub-discipline of the English literature/drama studies, but as a chief model of future Czech intellectual inquiry that is to steer the course of normative parameters of Czech cultural reality.

V. The collision of transitional Czech Shakespeares: 1964 and 1864

Aside from emphasizing Czech Shakespeare's intimate ties with the West – to counterbalance the powerful yoke to the East – many of the contributors link their immediate context of budding transition to a similar moment in Czech history where Shakespeare became a central pillar of nation-building. The nineteenth-century Czech National Revival proved an ideal resource. First, it provided obvious parallels of Czech national resistance to a colonial power. Second, as it foregrounded the struggles of peasants, the proletariat and progressive intelligentsia against a capitalist superpower, it had been incorporated by communist historiography at the core of the official Czech history narrative and thus posed as a safe subject of later celebration. Finally, it provided hope: the revival's teleological links to the eventual emancipation from the Austro-Hungarian Empire at the close of the First World War in 1918 suggested the potential for the twentieth-century Shakespeare quatrocentenary to upset the Soviet world order.

The Shakespeare tercentenary in 1864 was used to mark one of the pinnacles of the revival, second in importance perhaps only to an armed rebellion for autonomy of 1848. Ever since the first translation of *Macbeth* by K.H. Thám (1786), an ardent supporter of the Czech vernacular and author of *The Defense of the Czech Language*, Shakespeare had grown in the Czech cultural consciousness as a symbol of cultural achievement equated with eligibility to national independence. As Thám sought to raise the competitive national consciousness of his audience and to widen its internationally-recognizable cultural horizons, his translation yoked separatist anti-Habsburg politics with the cultural literacy of a critical mass of Czechs whose sensibilities Shakespeare would presumably groom towards appropriate universal ideals. Besides individual self-improvement, Thám's translation was to enable a collective experience of cultural worthiness in providing proof that the Czech tongue was linguistically capable of encompassing Shakespeare's genius. The Czech desire to appropriate Shakespeare was not only driven by a desire to claim cultural equality with other Western countries, but also to underscore Czech superiority over the rest of Eastern Europe. As Vilém Mathesius attests in his monumental multi-volume patriotic work *Co daly České země Evropě a lidstvu* (*Contributions of Czech Lands to Europe and Humanity*), Czechs have been invested in presenting themselves, paradoxically, as a model of Slavic nationhood and simultaneously as harbingers of Western culture. Such sentiments made it possible for Mathesius to claim that, from the 1860s, 'Southern Slavs gladly came to learn from the more evolved Czechs, and also gladly welcomed them in their homelands as tried pioneers of the nationalistic and Slavic spirit, as well as the carriers of the West-European cultural type' (1999: 92).

Shakespeare became a central lynchpin of the nation-building process of a small European country bent on proving its worthiness to its powerful neighbours, located predominantly to the west. His importance as a subversive political tool was underscored by the repressive response of the Austrian authorities to Thám's initial translation, which appeared contemporaneously with the unsettling upheavals of the French and North American Revolutions and in the midst of rising Czech cultural awareness. The translation was swiftly banned and all accessible copies were destroyed. This forceful censorship boosted growing cultural mobilization: the few remaining illegal copies became culturally precious, and those who kept them became known as subversive radicals. Not surprisingly, subsequent Shakespeare translations into Czech, which at this point was the language of the working masses and patriotic intelligentsia, did not fare much better. It was not until the mid-nineteenth century that new, unhindered and more thoroughly-edited translations of Shakespeare appeared alongside a proliferation of literature, poetry and philosophy written and published in the Czech vernacular. Crucially, these translations did not compete culturally with the original English versions, but rather sought to prove the Czech language's artistic superiority over German, in which Shakespeare had initially appeared in the Czech lands.

The rising need for visibility of Czech cultural consumption fuelled a call for appropriate venues, most prominently among them a theatre 'of our own'. The Shakespeare Jubilee in 1864 provided a crucial impetus for a nationwide fundraising effort to gather popular means for building a Czech National Theatre.[7] Contemporary nineteenth-century thinkers wrote about the centrality of both as markers of Czech cultural and political achievement, usually couched in terms of direct competition with the dominant German culture. Jan Neruda, a central literary and scholarly figure of the National Revival, underscores the interrelated importance of Czech theatre and politics, in which cultural achievement was synonymous with political eligibility for national independence. The availability of Czech theatre became a pressing nation-building necessity of nineteenth-century Prague, in large part to make room for Shakespeare plays in Czech translations so that the 'less corrupt' Czech audiences could hone their already superior understanding of the 'wide, calmly flowing, English humor' (Neruda, 1966, vol. 2: 23). Exposure to Shakespeare, Neruda wrote, would provide much more natural entertainment for the Czech audience than any 'sickly deformed portrayal of Viennese farce' (1966, vol. 3: 291) and, simultaneously, would enable Czechs finally to claim their share in the 'property of the educated world' (ibid.: 309). It is only through a national theatre, without which 'a culture does not deserve to call itself a nation'

(Neruda, 1918: 21), that the Czech nation can prove 'its moral strength, shape its thought, and let shine its passion for all that is good and beautiful' (ibid.: 3).[8]

The Shakespeare Jubilee of 1864 was conspicuously arranged to underscore Czech cultural commitment to Shakespeare as a display of Czech worthiness to cultural and political independence. Its various events sought to showcase Czech mastery of the playwright's work as well as suitable nationalistic – yet appropriately genteel – sensibilities that brought together the Czech spirit and particularly salient Shakespearean excerpts and characters. Overall, the Jubilee took on proportions of a political and cultural demonstration where representatives of Czech public, civic and artistic life pledged their allegiance to the Czech emancipatory nationalistic project via Shakespeare:

> Preceded by a cycle of five Shakespearian performances, the Jubilee itself was conceived as a composite work of art, consisting of music, poetry, singing, acting, painting, and sculpture. It took place in the largest theatre hall in central Prague ... The climax was reached when Smetana's[9] heroic *Shakespearian March*, freshly composed for the occasion, accompanied the procession of almost 250 characters from Shakespeare's plays dressed in historical costumes and slowly passing under a large bust of the poet created by the most promising young Czech sculptor of the day. The characters were represented not only by actors, some performing short scenes, but also by prominent patriotic citizens (Falstaff was impersonated by Prague's stoutest butcher). Finally the procession settled around Shakespeare's bust to listen to the poetic epilogue of the alleged bohemian shepherdess Perdita from the *Winter's Tale*. Represented by a popular actress, she embodies both Shakespeare's 'fair Bohemia' and bohemian art (Perdita Ars Bohemica) which was lost after the Hapsburg victory over the Bohemian estates in 1618 but was revived and fully restored now to become an equal partner of all European nations. (Stříbrný, 2000: 74)

The anti-colonial separatist rhetoric of nineteenth-century nation-building was well suited for appropriation by twentieth-century writers and authors who wished to express their protest against the colonization of the Czech lands by the USSR. Ostensibly celebrating the strength of the 'Czech spirit' in the face of a foreign colonizer in a not-so-distant past, these more contemporary nationalists harness the rhetoric of the nineteenth-century National Revival to agitate resistance against current colonial oppression. One among many, novelist Miroslav Slach places his *Usměvaví Rebelanti* (*The Good-natured Rebels*) (1959) in this time period. Based on nineteenth-century sources and closely following the conventions of the nineteenth-century Czech consciousness-raising novel, Slach's *Rebelanti* celebrates Czech culture in its traditional setting of a small regional town and its folk traditions. The novel serves both as a celebration of the peasant and working-class past that could well be endorsed by the communist regime and, in its focus on Czech traditions as a site of resistance to a foreign colonizer, as an indirect critique of the communist Russification of the nation. As theatre and print were central to the Rebirth movement, Slach's novel also speaks to the importance of Shakespeare translation and its later production for the forming of the Czech national consciousness. The novel's main focus is the creation of original Czech plays that would reawaken national consciousness in the audience, setting up a satirical mirror to reveal moral depravity of Germanized society and the hypocrisy of the upper classes. Shakespeare's work serves as a central touchstone in this process. In several key passages, Shakespeare steps into the foreground as the basic pillar and role model for Czech nationalism.

In Slach's patriotic historical novel, the connection between Shakespeare and the budding Czech national identity becomes explicit in a survey of the most important published works of the period. Martin, the town's aspiring schoolmaster dedicated to the nationalistic enlightenment of his predominantly poor pupils, visits the local pastor's library which, for its time, is unusually well-stocked with seminal texts of Czech and Western origins. Next to two nationalistic tomes about the Czech question, a Czech history by Pelcl (*Nová kronika česká*) and a monograph, ironically in German, on the history of the Czech language and literature by Dobrovský (*Geschichte der bőhmishen Sprache und Literatur*), Martin finds 'a small volume – Shakespeare's *Macbeth* translated by Karel Ignác Thám in 1786 – and that particularly caught Martin's heart' (Slach, 1959: 90). Even though the list continues, no other patriotic, philosophical or literary work fascinates him as much as the combination of the three. He becomes literally 'breathless' when he is given this 'first translation of Shakespeare into Czech ... the first jewel of an endless treasure' (ibid.: 94). Shakespeare in Czech, being the 'first' jewel among the others, clearly surpasses even the patriotic works in importance for the education of the working classes and the awakening of the nationalistic spirit mainly by the mere capability of the Czech language to encompass his works.

Slach's plot hinges on a literal competition between Czech theatre, staged by visiting university students at the local alehouse, and German theatre commissioned by the local gentry to be performed at their country seat. To put the Czechs in their place, the gentry chooses Shakespeare's *A Midsummer Night's Dream* where the fifth-act rustics' theatrical blunders are to ridicule the attempts of the 'lowly' students to perform theatre (ibid.: 109). The performance itself, however, shows that 'immortal Shakespeare' works in favour of the patriots, as the competition between the foreign gentry and Czech students awakens further curiosity of the general population in the surrounding region. The students' performance of a new Czech satire on the Germanization of the Czech population becomes an overwhelming success both as a play and as an incentive for de-Germanization at the grassroots level in the town and surrounding villages. Shakespeare here literally paves the way for a larger revival movement in shaming the middle class into renouncing their Germanized ways while buttressing the importance of the Czech nationalistic spirit.

Similarly to Slach's *Rebelanti*, the contributors to *The Charles University on Shakespeare* cheer on the National Revival as a moment in Czech history when Czechs successfully rallied to win autonomy from an oppressive system to free the nation from the rule of foreigners. This 'interest' in history, though crucial to the understanding of the process of Czech nation-building and cultural production, played a secondary role to the importance of this moment as a parallel to the communist-occupied Czech present. The explicit celebration of nineteenth-century Czech Shakespeare thus became not only an implicit critique of the current mid-twentieth-century political situation, but also an invitation to free the Czech culture from suffocating Russian influences.

A representative example is the contribution of Jan Mukařovský, who calls on Shakespeare 'as a political force ... so characteristic for the Czech theatre in the fight for national liberation' (1966: 12).[10] In the outlines of his celebration of the revivalist Shakespeare a century earlier, we can trace the blueprint of resistance available in the current sociopolitical moment. The subversive possibilities offered by Shakespeare in the nineteenth century are startlingly similar to the possibilities outlined by Mukařovský's colleagues, who focus explicitly on the minutiae of reformist Shakespearean interpretation in the 1960s. First, Shakespeare promises access to an alternative source of cultural power to the enforced culture of the current dominant empire. Second, his works reportedly serve as a repository of alternative models of moral and social structures. Third, and finally, his cultural capital provides a source of identity-building for citizenry worthy of imperial emancipation. In his characteristic celebration of the nineteenth-century Shakespeare Jubilee and the wider National Revival, then, Mukařovský reaches beyond a singular ideological discourse of Soviet socialist realism to a wider multiplicity of viewpoints resulting from 'direct contact with the broad world culture' (ibid.: 15). Eventually, this broader view of Shakespeare and the world is to cohere in a national 'Czech culture [free] from overwhelming foreign influences' (ibid.: 12).

Like many of the nineteenth-century thinkers, Mukařovský sees the greatest importance of Shakespeare in the inherent ability of his texts to 'improve' individual readers and viewers and who, armed with Shakespeare's code of ethics and political engagement, become better equipped to participate in dismantling the totalitarian public sphere. A mere exposure to Shakespeare's plays teaches Czechs the difference between 'good' and 'bad' art and refines their sensibilities in demanding appropriate dignified conditions of political existence in relation to the dominant imperial power. The 'general level of cultural education' (ibid.: 16) that Shakespeare thus provides as an alternative to totalitarian ideology becomes central to movements of nationalistic resistance that could be tapped in the present anti-communist moment.

As much as the tercentenary of Shakespeare's birth in 1864, the quatercentenary celebrations in 1964 took on explicit political overtones aimed at emancipation from the hegemony of Soviet imperial domination. The 'no end of Shakespeare in Czechoslovakia' that Stříbrný proudly reported at the Folger Library in 1964 included countless performances, lectures, translations, radio programmes, publications, international visits by Shakespeare dignitaries and theatre troupes, and, following the template of the tercentenary, another direct address to the playwright from the Czech nation (penned by Jan Werich, a widely-respected comic actor known for subversive political stage acts) (Stříbrný, 2007: 174). The collective celebrations no doubt participated in the unfolding reformist movement, abruptly ended in the summer of 1968. Yet, as the Soviet tanks rolled through the Czech countryside and into Prague, the Shakespearean discourse of transcendental freedom was revealed to be mere hopeful rhetoric. Despite the intellectuals' claims of common Shakespearean humanity with the wider 'free world', the 'free world' did little to stem the Red Army's mission to rail in the Czech culture to the hegemonic fold of Moscow-controlled communist ideology.

VI. Czech Shakespeare under normalization and 'real' socialism

Given the traditional centrality of theatre to Czech popular political life, it is not surprising that Shakespeare productions in general protested against the Soviet invasion in 1968, after which, in turn, the surveillance of artists and scholars intensified. Though the political processes to silence the most prominent and 'dangerous' thinkers no longer ended with executions, many intellectuals were fired from their jobs and were reassigned to work in factories and as street sweepers. Some were imprisoned, while others were effectively silenced. According to historians Čornej and Pokorný, 'the dissenters were not allowed to work in prestigious posts. The state, as the monopolistic employer, allowed them to work only in sectors where they had no contact with other people ... Children were often punished for the trespasses of their parents, in that they were barred entrance into higher education ... the reaction of the population was a growing apathy' (2000: 77). Some of the active cultural production moved outside the country through Samizdat and Western presses, laying the ground for the later popularity of artists such as Václav Havel, Milan Kundera or Bohumil Hrabal.

Most Czech artists, however, compromised and acclimatized to the new political regime in order to continue their creative work, occasionally expressing support for communist ideology with the help of varying degrees of official coercion. In the initial stages of normalization and real socialism, artists who depended on wide audiences to disseminate their work were, as Milan Otáhal has shown in his study of the interrelation of Czech intelligentsia and politics, unwilling to sacrifice their immediate and intimate access to wide swathes of the population (1999: 23). On the whole committed to eroding the ideological hegemony of the state through small acts of ideological subversion in their artistic media, they were far less likely to participate in open political protest that could result in their 'reassignment' to a different sector of the workforce. Dissenting writers, painters and sculptors were on the whole far more ready to express their discontent, relying on the alternative dissemination of their work to limited exclusive audiences. Only a few of the 250 signatories of Charta 77, a radical petition presented to the communist authorities on 1 January 1977, which called for state attention and adherence to universal human rights and for keeping international agreements, were theatre professionals.

The involvement of theatre personnel in open dissent gradually increased in the 1980s, in parallel to the increasing economic crisis within the USSR and the subsequent reformist measures of the perestroika inaugurated by Gorbachev. Taking advantage of the loosening grip of centralized communist power from Moscow, an increasingly visible underground of dissenters began to surface and grow, further weakening the state apparatus that was increasingly incapable of stopping popular opposition. Czech theatre professionals intensified their efforts to express viewpoints alternative to the officially condoned ideology and actors began shifting towards open protest. As in all turbulent moments in Czech history, theatres became popular meeting places where audiences and actors engaged in an intense, increasingly explicit political exchange, and theatre attendance grew exponentially. In 1987, theatre professionals lobbied for a new reform of the legislation on public theatres, asking for a reduction in censorship and more artistic autonomy (Otáhal, 1999: 27, n. 40). In support, a prominent Czech actor, Miloš Kopecký, called for a regime change, arguing for the need for complete freedom of expression for artists and particularly actors. In the same year, National Theatre actors entertained their audience with explicit political references in Shakespeare's *Love's Labours Lost*. For the first time actors on a communist-controlled Czech stage – and a stage of the National Theatre at that – dared to defy an official decree that the group of aristocratic youths, who in the play disguise themselves as 'visiting' Russians to woo their ladies, mask as 'Persians' instead. Thanks to this defiance, crowds of delighted Czechs could for the first time laugh at a public rendition of bumbling and thoroughly ridiculed pseudo-Russians who proclaimed to bring 'nothing but peace and gentle visitation'.[11]

The connection of theatre and politics became the most intimate and intense at the time of the Velvet Revolution, unmasking not only the workings of the stage as a central site for political debate, but also the trust of the population in artists, and actors in particular, to communicate facts concealed by communist authorities. Like previous historical moments of political crisis and change, by 1989, theatres played a crucial role in the growing social unrest against the soon-to-fall communist regime. The arrest of Václav Havel, a playwright, in January 1989 mobilized the growing political underground of citizens stepping into the arena of open defiance of the communist system, and the list of signatures on various political petitions grew exponentially. After a student-organized manifestation of popular dissent on 17 November,[12] which began the so-called Velvet Revolution, political meetings were held in theatres in the capital and around the country. The students of DAMU (the academy for dramatic art training actors, directors and artistic staff) and actors were the first to issue an official statement in protest against the widely-publicized police brutality of 17 November, calling for a general strike of all intellectuals and later the entire nation. The strike itself was first announced at Realistické Divadlo (the Realistic Theatre) in Prague on 18 November. The first public gathering of dissenters organizing the next stages of the 'revolution' took place in another theatre venue, Činoherní Klub. The first press conference was held at Laterna Magika, a theatre aptly called the Magic Lantern (Otáhal, 1999: 138–43). In the first days of the revolution, when the media still disseminated information of unsure origins and questionable trustworthiness, people either partook in daily protests in the city centre or gathered in theatres all around Prague for discussion and to organize the Občanské Fórum (the Civic Forum). This Forum provided an umbrella for individuals from varied backgrounds and turned out to be the platform as well as the moving force for the following political discussions and changes. The first president of the postcommunist Czechoslovakia, elected in the last days of 1989, was Václav Havel.

VII. Postcommunist Czech Shakespeare

One of the first challenges of the postcommunist era called for a return to a 'true' Czech economic, social, cultural and national identity, free from foreign interference. As elsewhere in CEE, the 1990s brought one of the first promising opportunities to deliberately define an independent modern state.[13] In the absence of the pillar of centralized repressive government, the postcommunist Czech culture looked to its pre-colonial past, its dissidents and the West for suitable models of both economic and cultural structures. The new independence brought an existential quandary, which some have called a 'cultural crisis', as neither the West nor the past offered suitable comprehensive cultural models, and the lived reality of the 'free', capitalist and nominally Western nation failed to bring the quick comprehensive recovery promised in return for the patient 'belt-tightening' asked of the citizens within the context of postcommunist neoliberal 'structural readjustment'. Though the material havoc of the neoliberal measures was not as devastating as elsewhere in CEE, largely thanks to the nationalistic – and in many ways ethnocentrically elitist – refusal of the Czech government to take exploitive loans from the IMF and World Bank (Wedel, 1998: 40),[14] the first decade after the fall of communism saw an unprecedented rise in unemployment, poverty, homelessness, crime and substance abuse. The economic boom visible in the affluent pockets of Prague and a few of the larger municipalities failed to reach all Czech citizens equally, who became polarized in terms of widening income gaps.

Dissident philosophies, which for four decades watched over 'true' cultural roots of the repressed nations, also proved less than cohesive as the revolutions of 1989 removed the stable ideological anchor against which widespread cultures of dissent previously defined themselves, glued together by anti-communist solidarity. New, more 'truthful' societal organizations lauded and propagated by anti-communist dissent, like Václav Havel's widely celebrated 'civil society', proved to be idealist – and even elitist – utopias, useless in the face of current realities without the resented yet necessary crutch of a socialist political system that provided its ideological – as well as material – departure point. Rather than a swift return to true roots, the process proved to be one of complex ideological reinvention made more complicated by the dogged determination to erase its blueprint lines that would challenge its claims to cultural authenticity. This process, moreover, became increasingly harangued by the EU which, though nominally committed to national independence, exerted significant pressure towards political, economic, cultural and subsequently ideological hegemony.

The heart of this process, located in broadly defined 'culture' and exemplified in artistic cultural production, equally lost the stable communist pivot around which – for better or worse – all previous artistic production turned in varying degrees of assimilation, subversion and resistance for more than two centuries. In the years immediately following the fall of communism, most theatres, for instance, also experienced a temporary identity crisis. Quickly shedding the repertoire recommended – or at least approved – by communist cultural advisers who worked to ensure ideological hegemony in expressive art, theatres sought new material. Instead of the dreamed-of freedom of artistic pursuit, the postcommunist adjustments, which frequently resulted in an unceremonious withdrawal of public funds from most venues accompanied by a cessation of state-sponsored bussing of rural audiences to approved performances, led towards constraints of cultural consumerism where art became only as viable as the tickets it could sell or the sponsors it could draw.

The postcommunist scramble of the Czech theatres for a new artistic identity that would sell enough tickets and bring commercial sponsorship effectively marked a complex shift in ideological control. Though theatres under communism operated under the close supervision of the state, which tempted with a financial carrot of production costs, salaries and artistic investment, while threatening with a whip of political repression, most theatrical productions aimed to exploit the multivalence of theatrical performance to subvert official communist ideology. Thus, many communist-overseen productions at least partially catered to the fairly unified subversive tastes of audiences who frequented theatres precisely to experience moments of communal ideological revolt. The fall of communism did not immediately diminish the habits of the theatre-going public, but it shifted the performative core from revealing the flaws of communist ideology to a much more ambivalent re-enactment of 'freed' culture by the mid-1990s. The policing of the state was replaced by self-policing of artistic expression realigned with market forces. Startled theatres aimed at the new postcommunist tastes of the admission-paying audiences in productions that could be marketed to potential commercial sponsors who were even more attuned to the solvent potential of the cultural capital of ideologically astute artistic expression. Rather than experimenting with an unanchored, wide-ranging new freedom of expression, theatres had to come to grips with the notion, articulated well by Michael Bristol, that 'cultural production is an apparatus of social regulation and control for the system of industrial capitalism as a whole' (1996: 114), without appearing to sacrifice art to a new – now capitalist – ideological construct. It could be surmised that, in this deliberately staged marketing moment of learning to play to the voracious yet undefined audience tastes in the first decade after the fall of communism, theatre became subtly reflective of the shifts, tensions and debates in the larger culture. Tellingly, at this intersection, most venues went in one of two fundamental directions: previously forbidden material, ranging from dissident plays to contemporary Western pieces, and tried box-office successes.

Among the latter, Shakespeare was enlisted as a surefire artistic heavyweight and, alongside a few other works positioned dead-centre in the traditional Western theatrical canon, was to fill the repertory gap created by the postcommunist confusion. While the theatres that reached for contemporary material were few – and even fewer dared to prepare new contemporary material produced locally – the likes of Molliere, Mozart and Chekhov, alongside Czech playwrights who came to the forefront in the nineteenth-century National Revival, sold out tickets to audiences trained to connect regularly with the cultural base traditionally found on the stage. In all the years for which the Czech Theatre Institute gathered official production data in the postcommunist years, Shakespeare rose to the challenge admirably to lead by leagues in all of the categories measured: number of productions, performances, venues and paying customers, totalling, on average, about 10 per cent of all tickets sold annually.[15] A random sampling shows, for instance, that in a two-week period in May 2001, 13 theatres in Prague alone staged no fewer than 33 performances of 11 different Shakespeare plays, not counting two or three daily performances at the Czech Globe replica. In 1994, President Havel called for a cultural revitalization of the Prague castle. Agentura SCHOK promptly responded by inaugurating the annual Summer Shakespeare Festival (*Letní Shakespearovské*

Slavnosti) in one of the castle's renaissance courtyards. The Festival, undersigned by the President and sponsored by a number of prestigious and most solvent national and international corporations, grew from a run of two plays over the course of the summer to presenting (since 1999) visiting Shakespeare productions from the nation and the rest of Europe; moreover, the Festival spread to several additional locales in the Republic and in Slovakia.[16]

Though the fall of communism did not diminish the number of Shakespeare productions in the Czech Republic, the centre of gravity of the performances shifted. The idealized universal Shakespeare, which under communism often served as an unquestioned subversive link to the rest of the (Western) world, suddenly came face to face with the reality of the universe beyond newly opened national borders. Instead of finding its mirror image in Western Shakespeare productions and scholarship, it was confronted by a rather ambivalent, ambiguous, postmodern multitude that evoked only selectively – if at all – Shakespeare's universal genius. Czech Shakespeare, too, in the absence of repressive communist ideology, branched out into new theatrical interpretations that sought to capture the solvent imagination of postcommunist audiences with multiple tastes. True to the newly re-established practice of separating art from politics and fuelled by the notion that, in a free society, art floats free of pedestrian political concerns, these productions have been widely hailed – and sometimes critiqued – as aiming at entertainment, rather than engaging in a political debate.

Czech Shakespeare scholars affirm this welcome separation of culture and politics in the postcommunist era. Martin Hilský, for instance, argues that, after 1989, 'the political dimension of literature and of the theatre was lost. The huge change in cultural semantics in the postcommunist Czech community consists in the canceling out of that resonance that almost magically generated the political meanings in any literary or theatrical text'. Relegating a connection between art and politics to the repressive communist era, Hilský posits that 'the political consensus, the shared anti-totalitarian attitude, which was such a powerful source of meaning in the Czech theatre, was replaced by the plurality of voices; and one cannot rely on that unanimity of silent dissent that one always found in the Czech theatre before 1989' (1994: 158). Similarly, in *Shakespeare and Eastern Europe*, Stříbrný concurs that 'Shakespeare's plays and poems are now received with less political alertness: less as a matter of life and death and more as a thing of beauty and truth' (2000: 146). Yet, despite this depoliticization, Shakespeare's cultural capital is unshaken and 'he continues to hold his mirror up to nature, society, and each individual' (ibid.: 136). Both Hilský's and Stříbrný's comments underscore the undeniable material difference in postcommunist Shakespeare production that no longer needs to function as a mouthpiece of ideas alternative to a suffocating dominant ideology of the state regime. This separation seems analogous to a necessary divorce of a marriage of convenience, where art and dissident politics that had joined forces as a necessary counterbalance to a repressive public sphere no longer find pleasure in each other's company. While the singular and predictable connection has indeed dissolved, however, the two spheres are hardly estranged. Indeed, in order to remain relevant to postcommunist audiences, the cultural sphere inevitably has had to continue the ongoing discourse of presenting materials that paying customers would willingly consume, and so has been inevitably suffused in the current sociopolitical moment. In this context, ironically, even the overt denunciation of politics in the cultural sphere is evidence of the sphere's minute response to the ideological shift that dictates that the boundaries between art and politics be firmly drawn.

Such public renunciation of politics in Shakespeare as well as the multiplicity of new productions should not be mistaken for a total plurality or freedom of expression in Shakespeare. As in earlier transitions when Shakespeare came to the forefront of the cultural sphere, the multiplicity of Shakespeare productions continues in a largely bardolatrous attempt to articulate 'what Shakespeare meant' and how, subsequently, Shakespeare's meaning relate to the current historical moment. The multiplicity of readings, then, are not aimed at widening the general view of Shakespeare, art, reality or humanity, but rather at articulating a new singularity that would approximate the 'truth' that is to lie at the core of Czech postcommunist nationhood. As such, Shakespeare has increasingly been evoked as a site of budding resistance to another perceived imperial hegemony that has entered the Czech political consciousness together with the cultural requirements that need to be met prior to the accession of the country to the EU. In resisting another cultural hegemony that paternalistically presents itself as superior to the candidate state, Shakespeare again functions as the impartial transcendental repository of truth that can be aimed against the EU's condescending rhetoric. Shakespeare's nominal departure from the site of politics – and vice versa – then only strengthens the position of Shakespeare as an impartial commentator on the postcommunist context.

The nominal withdrawal of Shakespeare from politics further underscores the seemingly private locale of some of the most pronounced tensions in the postcommunist Czech culture. Instead of commenting on the general state of the nation, easily facilitated by tragedies such as *Hamlet*, the postcommunist focus on comedies and romances betrays the seemingly entertainment-oriented emphasis on the private spheres of individual, largely domestic relationships. It is possible, I admit, that at least initially this focus could have been incidental to the postcommunist artistic confusion and symbolic of a common denominator of general human concerns at a time of relative peace. Yet, as the 1990s progressed, the political dimensions of such domestic focus became inescapable as its parameters came to dominate political debates in the Czech Parliament, where democratically elected representatives of the populace repeatedly and contentiously took up questions of normative boundaries of Czech citizenship in response to EU stipulations of 'human rights'. These debates not only addressed the formal vestiges of national belonging (such as immigration) but delved into the more intimate normative parameters of racial and gender relations, homosexuality and structures of social services that distribute material benefits to satisfactorily-performing citizen subjects. While such debates, particularly when followed by appropriate legislation, propelled the Czech Republic towards membership of the EU, average citizens, all of whom were likely to be affected by a genuine shift in normative parameters of the domestic sphere, began seeing a resurgent state control that sought to delineate daily behaviour and human interaction. In this context, the focus of Shakespeare productions on the interpersonal aspects of his plays came to participate in the wider political consciousness of the increasingly disillusioned – and largely resistant – populace.

The divide between the perception of the public and the private spheres is palpable in the public reaction to many of the new laws. For instance, in the public sphere, to avoid investigation by the authorities, employers have had to consider seriously raising women's salaries to match those earned by men, so that official company records in the near future begin reflecting 'equal pay for equal work'. Sexual harassment, in contrast, though highlighted as a major issue in the media, has been seen largely as a ridiculous complaint of overprivileged American housewives, and it has become the butt of many jokes, effectively silencing many who may find the equalizing measures appealing. In the private sphere, despite recent publicity, domestic abuse remains alarmingly widespread, condoned not only by the population at large, but also the authorities – both in the police force and in the medical field – who seem reluctant to respond to pleas for help beyond treating the victim's most grievous injuries. The victim's domestic situation is a matter of her or his private life. The state and those in the victim's most immediate environment are hesitant to become involved.

In the field of sexuality and race, the situation in the past decade has similarly pointed out the reluctance to accept these issues as political, despite numerous discussions about them in political and public venues. Publicly, homosexuality has been decriminalized and de-pathologized, and the senate has already discussed three times in the past decade legalizing domestic partnership and same-sex-parent adoption. Nevertheless, the homosexual population is to remain in their private domestic sphere. It is invisible outside its designated bars and clubs, as the public display of same-sex affection is seen as distastefully transgressive and is often 'punished' in outbursts of random street violence. Parents still routinely send their children to sexologists for diagnosis of the degree of their homosexuality and to learn the likelihood of a cure. Similarly, at the bidding of the EU, which has deemed the widespread racial discrimination of the Roma population as the gravest barrier to the Czech Republic's successful candidacy, the Czech government has begun a sweeping campaign of public education and legislation to ensure rapid shifts both in the social system to support the ethnic minority and in the prevailing racist attitudes in the population at large. Despite these efforts, violent crime against minorities has remained widespread and opinion polls from the late 1990s showed that racial prejudice has not diminished.

What counts as political is itself a profoundly political issue. The investment in continuing to perceive the issues of sexuality, race and gender as private effectively forecloses any possibility of a systemic change that would threaten the present status quo by allowing for a wide-ranging shift both in policy, response and public opinion. More importantly, however, since the existing national habits are part and parcel of a Czech tradition, the investment in de-politicizing these issues is parallel to a commitment to a traditional Czech identity. The official changes in normative boundaries of Czech nationhood that have entered many dimensions of the Czech economy and central politics are only ambivalently welcomed into the sphere of Czech privacy, for that is where the seeds Czech nationalistic spirit and identity have been kept even in times of the most profound cultural assault by the many colonizers of the Czech lands. The larger ambiguity about the future within the EU consists of, on the one hand, a desire for belonging in the privileged West and, on the other hand, a fear of loss of the Czech national identity in favour of a bland consumerist state operating on the basis of supply and demand, and controlled from abroad through EU laws legislated outside of the Czech jurisdiction. In this context, the private is ferociously protected in private.

It is not surprising, then, that the turn of Shakespeare productions to entertainment, with a larger focus on the issues of love, romance and occasionally death in Shakespeare's comedies, romances and a few select tragedies, has been widely perceived as de-politicizing the bard and his works. Yet, if we consider that the drama that surrounds love and romance – in reality as well as its fictional depiction (and not only in Shakespeare) – is always driven by sexual desire, social, racial and class identity of the potential or existing lovers, the implicit political connotations become clear. As Part II of this book will demonstrate, the tensions attendant on the rearticulation of postcommunist subjectivity are projected onto – and transformed within – the cultural sphere of supposedly transcendental Shakespeare, evoked once again to counterbalance imperialistic dimensions of the dominant EU cultural ideology.

Part II

3

Translation Wars: Redefining Shakespeare in the Postcommunist Czech Republic

One correct translation of Hamlet *does not and will not ever exist, in the same way that there is not and will not be an only 'correct' interpretation of this great and mysterious play. Every new translation of* Hamlet, *similarly to its every new interpretation, is the seeking of something that can never be fully discovered. Still, this process itself has a meaning, and I would argue that this non-translatability and non-interpretability of Hamlet is the catalyst of the never-ending process of discovery of this ever-exciting text.*

Martin Hilský, 2000

It is not my intent to argue that new translations are the reason for all the absurd, and I think unfortunate productions, but they certainly open the door to them and enable them. If Shakespeare belongs in the history of our literature, theater, music and film as an extremely important and inspirational figure, any deformations in the transfer from English to Czech are the basis not only for grief … but because they create a misleading image of this genius loved and worshipped by generations, they are a forgery of Czech tradition and treason against our national culture.

Břetislav Hodek, 2001

In [Shakespeare's] works, 'a' is 'a', and 'b' is 'b'.

Jitka Sloupová, 2000

In spring 2001, the front page of the culture section of a major national Czech newspaper, *Lidové Noviny*, featured an article in which a venerable Czech translator Břetislav Hodek charged his prominent colleague Martin Hilský with 'betraying the Czech national culture' by making Shakespeare too accessible.[1] A public exchange between the two quickly ensued, joined by journalists and literary scholars who compiled textual, intellectual and ideological arguments in defence of one or the other's more 'true' Shakespeare. As surprising as the public nature of this exchange might seem, Hodek was not the first to publicly challenge Hilský's popular translations, used by an overwhelming majority of proliferating postcommunist Prague Shakespeare productions. A year earlier, an eminent theatre journal, *Svět a Divadlo* (*SAD*), serially published an unfolding exchange between Hilský and a respected theatre critic and translator of contemporary literature, Jitka Sloupová. Sloupová had objected to Hilský's translations of Shakespeare's *Hamlet* on the grounds of excessive elitism, which reportedly presented Shakespeare as 'a sphinx understandable only to the chosen few' (Sloupová, 2001: 161). Though seemingly on the opposite ends of the spectrum of 'elitism' – one arguing that Hilský's work is too accessible to the general populace, the other that it is not accessible enough – both Hodek and Sloupová presented a united front in suggesting that Hilský's understanding of Shakespeare exhibited grievous flaws that needed to be exposed to public scrutiny in the interests of the intellectual and cultural health of the post-communist Czech nation.

A disagreement about the accuracy of a translation is common enough in the field of translation studies, which is long accustomed to thorough debates about translational fidelity in a context where smooth linguistic, conceptual and contextual transfer from one cultural context to another is hard – if not impossible – to find. As Michael Cronin has noted, the issue of accuracy has been the central lynchpin of all translation debates: 'if translation has been faithful to anything over the centuries, it has been to the notion of fidelity itself' (2003: 68). And yet faithfulness in translation has been a thorny concept, deliberately disguising a perpetual jockeying of various simulacra of the source text for consumption in the target cultural context. The victorious simulation largely depends on the configuration of the comparative political prominence of the source and target culture (and relevant nation states), and the cultural capital of the source text in question, as well as the ingrained cultural expectation on the consumption of 'foreign' texts in the target culture. As Lawrence Venuti has demonstrated, it is generally expected that the few translations of non-English-language literature prepared for consumption in the English-speaking world appear in various guises of simplified exotic accessibility that implicitly underscore the Western ideology of intellectual superiority. In contrast, non-English speakers – particularly those outside the boundaries of the established West – have come to expect much more intricate renditions of English-language classics that foreground the values of the source culture for their presumed cultural edification.

Venuti has memorably argued that translation is a 'cultural political practice, constructing or critiquing ideologically-stamped identities for foreign cultures, affirming or transgressing discursive values and institutional limits in the receiving culture' (2002: 15). The very act of linguistic translation helps exemplify the heart of national identity, which, as Cronin reminds us, is 'constructed on the ground of [intercultural] difference' (2003: 90). This is not merely because 'symbolic expressions of identity, relations and history are powerfully, though not exclusively, expressed through language' (ibid.: 100); more importantly, any debate about the 'fidelity' of a translation to a source text invokes the necessity of affirmation of this difference as well as degrees of assimilation of one culture (in)to another. The wavering between the poles of domestication and foreignization of a text's translation – particularly when the aforementioned text comes packaged in layers of domestic and international cultural capital – evokes contextual cultural anxieties about forced, inevitable or merely accidental assimilation to worldviews that might eventually undermine the existential self-interests of the target culture's nationals. In a context where the target culture in question faces all-too-real pervasive attempts of Western powers to align its postcommunist cultural values with the central consumerist tenets of neoliberal capitalism, worries about the culture's agency of self-determination proliferate.

Even as the most recent Czech Shakespeare translation debate treads well-worn ruts between the two poles of acculturation and foreignization, it simultaneously makes visible the implicit outlines of the political beast underneath that engages Shakespeare cultural capital for the postcommunist nation-building project. Evoking warfare via imagery of national betrayal and treason, these debates underscore powerful cultural undercurrents that have surfaced in the ongoing public discourse of the postcommunist transitional process from communist totalitarianism towards the promise of a free, democratic state. The vitriolic passion of this disagreement paradoxically foregrounds the foregone agreement that the Czech postcommunist future is far from certain, teetering between a forcefully discarded totalitarian past, pressures of the West eager to profit from new economic and political frontiers, and internal political turmoil. Equally, the contributors to the debate unquestioningly assume that, in the context of creating a successful national future, Shakespeare *matters*; indeed, the continual debate about the translation of Shakespeare reinforces the centrality of Shakespearean cultural capital to a successful postcommunist Czech nationhood. The shades of disagreement merely make visible the otherwise implicit competition for the most appropriate simulacrum of Shakespeare for consumption in the postcommunist Czech cultural context; in tandem, this competition seeks to renegotiate the comparative political position of the postcommunist Czech nation in the context of unevenly profitable globalization. As such, the debate harnesses the disagreement over accuracy of words on a page to address the far more important question of the nature of Shakespearean cultural capital that is to assist in articulating a viable Czech postcommunist future.

Despite the wide-ranging implications of the debate, its publicized content revolves around predictable questions of content, linguistic transfer from one language to another and the equivalency of two disparate cultural contexts. Most obviously, it manifests a linguistic disagreement about translation of a Renaissance source text, produced in a specific Elizabethan context, into a twenty-first century East-European target culture. It equally re-enacts the faultlines of generational divides, in that Hodek champions an established traditional mode of translation guarded under communism as a repository of cultural expression alternative to communist ideology; in contrast, Sloupová champions the radical translations of Jiří Josek that are explicitly aimed at the Czech postcommunist youth; and Hilský argues for a heteroglossia of Shakespeare translations to better address multifaceted postcommunist transitionality. Additionally, the debate evokes a host of issues particular to its sociopolitical context. For instance, the disagreement also reflects an unarticulated – and in polite circles unarticulable – new material dimension of cultural production coalescing after the fall of communism. In disputing the merits of Hilský's work, both challengers refract a dismay at the channeling of scarce material funds into one set of (unworthy, they would have the public believe) hands, while alternative translations (his own in Hodek's case, and Josek's in Sloupová's case) are largely unnoticed. The litany of alleged inaccuracies in Hilský's translations is therefore not only aimed at his work, but also at the developing cultural consumerist framework that seeks to create public tastes via the marketing of a pre-selected set of cultural artifacts, while channeling material investment into the production of cultural consumables likely to satisfy the fickle appetites of the postcommunist, increasingly consumerist public. In disputing Hilský's popular translations' quality, then, the challengers protest the selective restructuring of cultural values that should operate, they suggest, on the basis of indisputable merit and intrinsic value.

As such, each version of Shakespeare emergent in these serial debates underscores its author's framework of contemporary cultural reference. Despite the pronounced clash of differences at the centre of the translation debates, all three participants demonstrate ardent commitment to Czech independence from foreign intervention in the transitionality of postcommunist nation-building. Equally, all three claim Shakespeare's transcendental status as central to securing this independence, though the precise dimensions of this transcendence vary according to the dimensions of cultural 'true' values that are to serve as a blueprint of postcommunist Czech society. Thus, Hodek's insistence on Shakespeare's inherent inaccessibility to the average citizen seeks to protect the 'transcendental' cultural core of the Czech nation from the perceived threat of neoliberal consumerist practices that would transform them into simplified, powerless, casual consumables. In contrast, Sloupová's insistence on the core necessity of Shakespeare's accessibility to all Czech citizens reveals a populist use of Shakespeare's assumed value as a democratic cultural vernacular through which the Czech masses can collectively stake a claim to their democratic, independent future. Finally, Hilský's pervasive insistence – in response to challenges from both camps – on the timelessness of Shakespeare inherent in the malleability of his texts to any cultural framework locates the political promise of Shakespeare in his versatility as an ideological tool that could be mined with flexibility for a variety of emerging socio-political needs.

In the rest of this chapter, I will consider in detail the specific dimensions of each translation argument to flesh out further each debater's position on Shakespeare and, subsequently, the future of Czech national development. In their criticism of Hilský's work, Hodek and Sloupová amassed a dense net of arguments connecting Hilský's translations to the disturbing trends of both popularizing previously exclusive art and making it more elite. Though they hypothetically dispute accurate translation based on a literal word-for-word transfer in contrast to the translation of meanings, the debates take various shapes in support of this charge. They range from definitions of the ideal scholar of Shakespeare to debates of historical positioning of Shakespeare's texts, discussions of accessibility of original and translated texts which lead to literal word-for-word discussions of particular passages, and competition over who 'loves and respects' Shakespeare more and has the more 'appropriate' attitude towards his works. The outcome is a complex – and ultimately untenable – web of criteria according to which Shakespeare can be claimed and reclaimed, as well as defined and redefined, by particular cultural factions and scholarly individuals. The stakes of this debate are high, revolving around the prize of intellectual agency to define the pinnacle of Czech cultural achievement and, more broadly, the heart of the Czech postcommunist nation and national identity.

I. Břetislav Hodek: foreignizing the present

The explicit goal of Hodek's challenge of Hilsky's work is to re-establish Shakespeare as a mysterious and ageless icon of humanity whose very value depends on the inability of the current age and time to comprehend fully all of its assumed complexities. Shakespearean cultural capital is thus simultaneously safeguarded from the temporal pressures of the present age even as it provides a bottomless repository of available models of human cultural interactions to be presented to eager masses by the few well-informed scholars of the inner Shakespearean circle. Such Shakespeare then appears shielded as much from the postcommunist political upheaval as from the pressures of Western powers – be it existing EU members, the US or the more ubiquitous Western media – to transform Czech society around capitalist neoliberal values that would assist Western economic and political interests. While this forceful removal of the source text from the target cultural context proposes Shakespeare as an artifact foreign to the contemporary Czech audience, Hodek's line of argumentation ardently maintains that it is *not* Shakespeare who is a stranger to Czech culture; rather, the Czech culture has estranged itself from the timeless and unchanging Shakespeare and all that he represents. The Czech tendency to claim Shakespeare as a founding intellectual further underscores this foreignization of the present: while Shakespeare remains unmoved, faithful to core human values, Czechs have moved into uncharted – and, Hodek suggests, untenable – sociopolitical waters.

Hodek's position presents an uncomfortable paradox: to claim one true and exact Czech version of Shakespeare negates the celebrated multiplicity of interpretive possibilities that Shakespeare's texts offer up. Such whittling away of superficiality from weighty essence thus inevitably calls for the intervention of a skilful translator/interpreter – whose credential are beyond reproach – who is to delineate the definitive outlines of the one Czech Shakespeare and to confer authority on subsequent cultural critiques. Not surprisingly, Hodek presents himself as one – if not *the* one – Shakespeare acolyte well-suited to the task of postcommunist cultural formation. His public persona, in the critiques aimed at Hilský, seeks to embody the venerable position of a persuasive orator within the traditional Renaissance mode of address. Evoking Ciceronian conventions of securing 'favor of audience', Hodek follows Cicero's suggestion to seem to speak out 'reluctantly and under compulsion'.[2] Accordingly, his initial public critique of Hilský, indirect in its ostensible aim at current theatrical Shakespearean productions, is shaped in the form of a reluctant interview conducted only 'after a long period of diplomatic silence' when Hodek was 'persuaded' to share his views on 'how William Shakespeare's plays should be produced and into what form of Czech they should be translated'. Later these claims of legitimacy are strengthened by Ciceronian claims to 'merit, achievements, or reputable life', underscoring the length of Hodek's engagement with Shakespeare and his texts. In an opening statement to his second lengthy critique, Hodek reminds the readers that he was shortly to graduate from high school, having already studied 'the Stratford genius for four years', when Hilský was born.

The consequence, Hodek argues, is that 'Professor [Hilský] always already has a handicap in the fact that my experience is longer'. And yet, at the time of writing, as Hodek turned 78, Hilský celebrated his 60th birthday; both had established themselves thoroughly in the Czech intellectual and scholarly field, a fact that casts some doubt on Hilský's categorization as the inexperienced young punk of Hodek's challenge. Rather than literally categorizing Hilský as inexperienced, then, Hodek's evocation of wisdom and experience seeks to critique the lack of critical reflection on the part of younger postcommunist generations slowly drifting away from the transcendental core of Czech cultural values.

Hodek's rhetorical invocation of Renaissance oratory traditions mirrors his Shakespeare's removal to the distant inaccessible past. Hodek proposes that an *ideal* translation of Shakespeare could have been provided by Jan Ámos Komenský (1592–1670), Shakespeare's near-contemporary known internationally as Comenius. This prominent Bohemian scholar, who promoted the didactic dimensions of all literary works as tools for developing the moral and national consciousness of readers, would, Hodek argues, have been the ideal candidate for transposing the worth of Shakespeare into an equivalent historical Czech context. Since no such translation exists, Hodek proposes to preserve the 'patina of the original' by translating Shakespeare's works into seventeenth-century Czech. While the explicit thrust of the argument lies in distancing Shakespeare from the present, more politically significant is the elegant elusion of the historical difference between Renaissance England and seventeenth-century Bohemia, two geographical regions Hodek yokes together via an assumption of uniformity of historical development. Such pairing of the English Renaissance – the ideologically-powerful assumed cradle of modern Western culture – with the politically marginal seventeenth-century Czech lands (no less celebrated by the Czechs) enfolds the Czech culture with the precursors to the 'developed' West. The link between Shakespeare and Comenius, which marks not only their contemporaneity but implicitly establishes the two historical figures as intellectual equals, symbolically bridges the Czech and English cultures, suggesting not only common roots but also a common cultural potential that, even if temporarily disrupted and decelerated by forcefully applied communist rule, promises political and cultural independence in the postcommunist world.

Hodek further buttresses this historically anchored preeminence of the Czech culture via an imaginary contest for the most profound evidence of Shakespeare's cultural influence outside England, a technique frequently evoked, as we saw in Chapter 2, by Czech scholars wishing to assert the legitimacy of Czech sociopolitical independence from foreign political interference. Staging such a contest, Hodek asserted that 'Shakespeare had immense influence on the development of Czech culture, even more so than in the neighboring countries'. Thus, within the seemingly academic frame of a dispute over Shakespearean translation, Hodek claims Czech cultural precedence over its neighbours, both to the East and West. To the East, at a time when the competition about precedence in inclusion into the EU was at its greatest, such cultural 'victories' contributed to the list of cultural accomplishments that would create clear grounds for accession precedence while minimizing the patronizing requirements that the EU imposed on the 'backward' regions of CEE. To the West, Czech precedence with Shakespeare could serve as a reminder of cultural superiority that should warrant unmolested cultural development even within the exclusive and forceful established club of the EU. To wit, Hodek's argument did not merely reiterate Czech cultural worthiness, but served as a forceful reminder that such worthiness cannot be taken for granted. Indeed, it is only through the careful preservation of traditional modes of Shakespeare translation, interpretation and staging that the Czech national culture will maintain its lead in the symbolic competition with its CEE neighbours. Perceiving an acute threat to this continuity from contemporary modernized productions, exemplified in Hilský's translations, Hodek asks: 'Is this Shakespeare? Is this the author for whose sake Prague organized a marvelous parade one and a half centuries ago? Was it for these characters that Bedřich Smetana composed his celebratory march?'

In this pregnant comment, like many scholars who seek to evoke the essence of Czech national identity, Hodek invokes the widely idealized nineteenth-century Czech National Revival.[3] This reference not only evokes the sacrifices made by generations of Czech intellectuals who are credited with the creation of the cultural Czech base responsible for the successful revolt against the Austro-Hungarian Empire that eventually resulted in Czech national independence, but also suggests youthful ingratitude on the part of those who wish to tamper with the long-established tradition of respectful veneration of Shakespeare. Shakespeare, as Chapter 2 sought to illustrate, provided a central pillar of the nineteenth-century National Revival process, wherein translations of Shakespeare's works served as evidence both of Czech cultural achievement and the capability of the Czech 'common' or 'vulgar' language to communicate Shakespeare's lofty words and thoughts. In his critique of contemporary Shakespeare translations, Hodek laments the loss of this 'two-century-long stem of Shakespeare tradition in the Bohemian lands' when 'the genteel nobleness and all the colors of Shakespeare's dramatic poetry' were expressed 'through the beautiful, pure, sacred mother tongue', which is currently in danger of being 'forsaken'. Hilský, as both the purveyor of such new reinterpretation of Shakespeare's works and a symbol of mass cultural deterioration evidenced by the wide popularity of his work, is held responsible for the current 'embezzlement of Czech culture of translation'. Indeed, the perceived inaccuracies in Hilský's translations do not result from a lack of knowledge, but are ascribed to a far less forgivable 'carelessness' about Shakespeare and 'condescension' towards his timeless works that privileges Hilský's egoistic interest in 'his own creative attempts' over the jewel of timeless universal culture that could – if handled properly – propel the Czech culture towards an independent future within the Western European Union.

The distance between the core Czech culture and present postcommunist cultural developments comes forcefully to the forefront in Hodek's merciless critique of the national Josef Jungmann Prize Hilský received in 1999 for his translation of Shakespeare's *Sonnets*. Named after a prominent linguist, translator and intellectual activist of the nineteenth-century Czech National Revival, the award explicitly evokes the parameters of the artistic dimensions of postcommunist nation-building, rewarding skilled appropriations of the (generally) Western literary canon for Czech cultural purposes. The core of Hodek's critique addresses precisely this conflation of translation with nation-building in an assertion that Hilský's work unequivocally fails to satisfy the requirements of embodying the transcendental values presumably at the heart of the nation-building process. The reality that Hilský's work, which according to Hodek 'only a few years ago … would not even be accepted to compete', elicited so much attention as to receive the first prize suggests an uncomfortable shift in the national valuation of aesthetics that, in turn, redefines the parameters of the nation in the postcommunist era.

Paradoxically, Hodek makes visible the otherwise denied ideological dimensions of artistic endeavours in the postcommunist cultural context, wherein, despite the rhetoric of postcommunist freedom of expression, cultural production is nevertheless closely tethered to the multilayered ideological dimensions of the postcommunist public discourse. After all, Hodek is undeniably correct in asserting that, under communism, Hilský's *Sonnet* translation – historically the first to allow for the homoerotic dimensions of Shakespeare's poem cycle – would have likely earned the translator a vocational reassignment from his university post rather than a prestigious national prize. Yet, while the communists would have likely been explicit in citing the work's failure in embodying the central tenets of productive socialist realism, the only recourse of Hodek's postcommunist censorship is in the nebulous area of aesthetics that hinges on the precarious legitimacy of Hodek's long-term literary and translation experience. The ideological dimensions of Hodek's critical thrust remain largely under the surface, not daring to openly point out the implicit allegation of political motivation in awarding the prize to a translation literally on the cutting-edge of Czech cultural comfort (with homosexuality); such unprecedented inclusion of homosexuality, even if based on the source text, Hodek's critique implicitly suggests, is far too sensational to assist in appropriately charting the already precariously positioned Czech national and cultural future.

However, the most informative thrust of Hodek's public protest against Hilský's translation is encapsulated in the brief question of linking Shakespeare's text to the cultural capital of his perceived identity: 'Is this Shakespeare?' As quickly becomes clear, the very identity of Shakespeare here depends on the level of accessibility to the average audience member, *not* on the actual content of a translation. Hodek goes as far as to suggest that Shakespeare, once made accessible, ceases *to be* Shakespeare. 'What is done to Shakespeare today is terrible', he claims; attempts to make Shakespeare's text more accessible to the '*kind viewer* are ludicrous. That strips Shakespeare's rich language'. Hodek's *kind viewer*, as becomes clear elsewhere in his critique, becomes synonymous with a 'less perceptive viewer' who can only be truly satisfied with a simplified, necessarily inaccurate, popularized and vulgar translation or, as Hodek articulates it, 'by castrating poetry, [Hilský] degrades dialogues and retorts of [Shakespeare's] characters to make them understandable even to the less perceptive viewer'. The very traits that make Shakespeare accessible, Hodek argues, negate the characteristics that elevate him from the position of a mere author to that of a spiritual icon.

Shakespeare's inaccessibility in Hodek's context inadvertently resurrects boundaries of social class that divide possible Shakespearean audiences according to their level of cultural sophistication, tastes and education. A 'less perceptive viewer' is not necessarily a synonym for an inattentive audience member, but rather for a viewer lacking the level of education Hodek implies is necessary for the full comprehension (and enjoyment) of 'true' Shakespeare. Tactically disregarding the historical reality of public Elizabethan theatre that catered to socially wide-ranging audiences with varying degrees of formal education, Hodek perpetuates the view that Shakespeare's texts are inherently inaccessible, which should be reflected in any translation. Rather, his argument suggests, both the text and the performance should be fully accessible only to the studied viewer already familiar with the English literary Renaissance. Any attempt at making the text accessible runs the risk of creating 'populist translations', which are by definition synonymous with a voracious consumerism of postcommunist 'money-grabbing' theatres who are more interested in selling a product that can be consumed by paying audiences than disseminating a cultural artifact to the cultural elite groomed towards its reception. Accessibility and popularity here gain explicit class characteristics, as, in Hodek's view, accessible translations tend to 'vulgarize Shakespeare's language' not only in linguistic choices themselves, but also in bringing the social strata within the plays about 'two social classes/rungs down'. In short, accessible translations change the very nature of the translated Shakespeare and, inevitably, the very essence of Czech national culture traditionally located in the elite intellectual. That is how Hodek arrives at the conclusion that the work of 'these translators' equates to a dangerous and unpardonable 'treason against Czech national culture! Making Shakespeare into an author who can be handled quite arbitrarily is terrible'. There comes a point, Hodek argues, when 'we have to bang on the table and say: enough, gentlemen, not this way'.

While the form of Hilský's translation receives the bulk of Hodek's criticism, the fact that such translations have appeared on the main stage of the National Theatre adds cultural insult to the national injury. The spatial signification of the National Theatre stage, which in theory is to showcase the best that Czech cultural production has to offer, has been demerited by the unworthy performances of Shakespeare productions based on populist translations. Horrified by a recent production of *The Comedy of Errors* (based on a translation by Hilský), which he likened to a 'stumbling farce', Hodek claims that 'circus manners could entertain [him] in a ring, but in Stavovské Divadlo they – to use decent language – embitter [him]'.

Even the concrete examples of textual infidelities and inaccuracies on which Hodek's argument is built perpetuate the elitist categorization of humanity parallel with his implicit categorization of contemporary postcommunist Shakespeare audiences. Citing the offending translation of *Errors*, for instance, he begins to analyse linguistic variants in translating the multivalent Renaissance epithet 'peasant', used in the play by socially superior characters to categorize in derogatory terms those relegated to lower (though not explicitly agricultural) status. Hodek deems both variants offered by Hilský, '*blbec*' (one of many Czech epithets roughly equivalent to 'idiot') and '*sluha*' (servant), inaccurate within the particular contexts of their narrative moments in the play. Instead, Hodek defends his choice, '*venkovan*' (literally, 'country-man'), as one that is the closest in meaning in that it both reflects its agricultural and geographical connotations of the original word as well as the historically derogatory connotations of the word, a translation he 'wrongfully hoped would immediately occur to all anglicists'. Where Hilský's selections aimed at making explicit the class power-struggle by underscoring either the negative epithet itself ('*blbec*') or the subservient archaic social position ('*sluha*'), Hodek's choice ('*venkovan*') asks the audience to acknowledge a feudal worldview that categorically demotes all countryfolk to an inferior class of nitwits. Hodek's reservations about the aesthetic renditions of poetic language similarly upholds historical pretensions to poetic sensibility. Comparing a notoriously overpoetical translation of *Errors* by J.V. Sládek to that of Hilský, Hodek juxtaposes Sládek's translation of a 'star with a fiery train' as '*hvězda s vlečkou ohnivou*' (a poetic and metaphoric linguistic equivalent) to Hilský's pragmatic and scientifically unexcited '*kometa*' (a 'comet').

The loss of poetic diction, together with the presumed erasure of awe at the mention of a once-unfamiliar celestial body, leads Hodek to conclude that Hilský has ceased to pay attention to Shakespeare text and has begun to translate the (presumably inaccurate) 'footnotes of editors'.

In a world where all three debaters use Shakespearean translation as shorthand for a wide cultural critique, Hodek's turn to accessibility as the key lynchpin of his argument seems logical. While the distance between the source text and the target culture is a given for all three, Hodek's insistence on maintaining maximum distance between postcommunist Czech culture and Czech translation of Shakespeare serves a twofold purpose. First, the elision of differences between the source text and the proposed archaic translation effectively appropriates Shakespeare's cultural capital into the Czech cultural canon, contributing to a stable intellectual core on which contemporary postcommunist culture can build. Second, the distance itself provides a convenient source of critique of current postcommunist developments; evoking the assumed value of transcendental Shakespeare, the distance itself suggests that the postcommunist Czech culture has distanced itself from the core humanist values presumably embodied in the source text *and* the translation Hodek proposes. Easy accessibility to the text/translation would undermine both the distance and the attendant critique, providing an erroneous – according to Hodek – impression that all is well in the state of Bohemia.

II. Jitka Sloupová: domesticating Shakespeare for the young guard

Whereas Hodek argues for maximizing the distance between Shakespeare and contemporary postcommunist Czech culture, Jitka Sloupová seeks to grind this cultural distance to dust and to claim Shakespeare as an immediate participant in postcommunist cultural developments. In her article 'Every Czech is a Hamlet: Enjoying *Hamlet*', Sloupová criticizes Hilský's translation as one that obscures Shakespeare's meaning at the expense of clarity and accessibility to a contemporary postcommunist audience. Her English subtitle, 'Enjoying Hamlet', is explicitly borrowed from a US university webpage on which a professor of psychiatry, 'with a typical American positive approach', provides a close reading of *Hamlet* to make the play more understandable to students, to foster their enjoyment of it and simultaneously to prevent them from committing suicide.[4] The explicit evocation of a US (non-expert) resource of Shakespeare exegesis simultaneously exemplifies a level of anti-elitism (Shakespeare is elucidated by a psychiatrist, not an institutionally acclaimed Shakespeare scholar) and aligns Sloupová's argument with the universalizing influences of US culture pervasively felt in the postcommunist Czech context. While refraining from a wholesale acceptance of a psychiatric exegesis of Shakespeare, Sloupová embraces the 'positive' method of wide domestication and dissemination of Shakespeare's works that she deems necessary for the productive growth of young generations of Czech citizens. After all, if US students are allowed unmediated access to the symbolic cachet of Shakespeare's edifying greatness, she implies, Czech youngsters should be offered a comparable advantage to maintain the cultural and political competitiveness in the complex sociopolitical postcommunist circumstances dominated by US ideological interests.

Most relevant to the translation debate at hand is Sloupová's comparison of translations of *Hamlet* by Hilský and Jiří Josek, Hilský's colleague at the Charles University, who has produced acclaimed translations of Shakespeare popular in theatres outside Prague (particularly Ostrava and Hradec Králové) and has distinguished himself in providing translations for dubbing and subtitles of many English-language films, including an internationally acclaimed BBC Shakespeare series. Of the two translators, Sloupová claims that 'Josek's translation wins hands down in regards to lexical closeness and humor', particularly as it underscores the attractive and easily digestible image of 'Hamlet [a]s a contemporary student with a bohemian vocabulary' (2000: 10). Sloupová's criterion for arbitrating the comparison/competition seems to be accuracy of the translation in relation to the original language in proportion to the contemporary currency of the used language. Drawing on several detailed examples from the English source text and the two translations, she claims that the main difference between the two Czech versions is that:

> Josek's translation attests to a more meticulous penetration of the meaning of the original, his version is more unique, less dependent on older translations, it speaks to a larger theatre

> experience and imagination, it is clearer and simultaneously most emblematic, contemporary in its vocabulary, richer in layers of meaning, and more careful about their preservation. The rhythm of his verse is more natural and temperamental, and his humor juicier. Hilský is generally closer to Shakespeare's mannerism, but in its transfer he is less fortunate than in the case of his previous titles ... the published and produced version of his Hamlet evidences a few blunders, if not outright mistakes, attesting to superficial, hurried work. (Ibid.: 9)

Unlike Hodek, Sloupová evidently finds the greatest accuracy of translation in a successful approximation of signification from the source materials to current cultural contexts, which equal a 'more meticulous penetration of the meaning of the original'. Pointing to Josek's more extensive 'theatrical experience', Sloupová is evoking a new set of criteria for judging Czech versions of Shakespeare. In particular, Josek's style of translation, which deliberately evokes the brevity and efficiency of expression necessary for subtitled translation, is equated by Sloupová with greater accuracy in translation and, on a higher level, greater authority for a Shakespeare translator.

The most significant aspect of Sloupová's argument above is her insistence on the greater accuracy of Josek's text in that it is 'less dependent on older traditions', more 'contemporary in its vocabulary' and therefore 'more natural' than Hilský's version. The defining boundaries of 'nature' and of the contemporary moment are here blurred to the point of becoming synonymous. As such, the value of translation becomes relative to the exact moment of production, so that Shakespeare's timelessness is paradoxically encapsulated in each contemporary moment of Shakespear ean interpretation. In this instance, then, like Hodek's approach, Sloupová paradoxically conflates Josek's contemporary translation of Shakespeare with its Renaissance version to argue for accuracy that seems to erase historical and cultural differences between the two texts. For Sloupová, Shakespeare is always a contemporary. Yet it is important to note that it is *not* a current appropriation of Shakespeare that she argues for; if that were the case, it would be easy to understand her position as one that recognizes certain themes in Shakespeare as relevant across centuries. In contrast,

Sloupová's bid for the *accurate* Hamlet as 'a contemporary student with a bohemian vocabulary' suggests that she evokes Shakespeare as a literal contemporary.

Sloupová's version of faithfulness to the source material equally evokes the valuable cultural capital Shakespeare presumably disseminates among contemporary audiences. In other words, Sloupová's Shakespeare is as singular as Hodek's. To illustrate this stance, Sloupová explains that when it comes to accuracy of translation, '"a" is "a", and "b" is "b"' (2001: 161). This equation suggests that the meaning of Shakespeare's work is fixed, so that an 'a' of the source text corresponds with an 'a' in the target language and culture. The erasure of cultural and historical difference is evident: the meaning and cultural connotation of 'a' in the source text can, should and must correspond with an identically signifying 'a' in the target language, rather than with a multitude of close but never exact possibilities: an '**a**', 'à', 'â', 'ã', 'ä', 'å', 'ǻ', 'ả' or the perhaps more questionable 'A', 'Á', 'Â', 'Ã', 'Ä', 'Å', 'Ả' or even the possibly subversive 'α', '@', 'ǽ', 'Æ'. As much as Hodek assumes the possibility of a direct transfer from Renaissance English into Renaissance Czech, Sloupová presupposes the possibility of a direct transfer from the Renaissance original into contemporary Czech.

The impossibility of 'accurate' translation is perhaps best illustrated by the implicit, never directly articulated disagreement between Hodek's and Sloupová's versions of accurate Shakespeare, on which they both insist. Whereas Hodek criticizes Hilský for making Shakespeare accessible to contemporary viewers, Sloupová criticizes Hilský for not making Shakespeare accessible enough but rather suggesting to the 'lay' audience that 'Shakespeare is a sphinx understandable only to the chosen few' (Sloupová, 2000: 161). Rather than an inaccessible vault of precious insights, Sloupová's Shakespeare becomes a mass intellectual mobilizer accessible to all who care to wield his cultural capital into the Czech postcommunist future. Unlike Josek's work, which Sloupová upholds as an example of the 'actualization and de-baroquization of Shakespeare' (ibid.: 10), Hilský's 'hollow' text can be successful only, Sloupová insinuates, when it is heavily and 'appropriately' edited down, as it happens in one of the productions she recounts, so that it 'seems to be ... a product of the contemporary culture' (ibid.: 12). Otherwise, Hilský's work seems to be a liability to any production; in fact,

Sloupová blames much of the shortcomings of the National Theatre production of *Hamlet* precisely on Hilský's translation (ibid.: 13, 15). By the same token, Josek's 'excellent' translation accounts for a full 'half of the success' of a production at the National Theatre of Brno (ibid.: 19).

The full cultural implications of Sloupová's comparison of Hilský's and Josek's translations surface in her discussion of the implications of the success of the Brno production of *Hamlet* which, she argues, was that it is of a 'grand, so to say "English" style'. The nonchalant erasure of semantic distance between 'grand' and 'English' suggests another instance of conflation of the Czech cultural achievement, when properly presented, with that of the Shakespearean symbolically and politically dominant cradle, England. The erasure of national boundaries, evocative of the impending accession to the EU, suggest that the Czech Brno Hamlet as a 'contemporary student' would fit equally a Czech classroom *and* an English one, erasing sociopolitical differences between contemporary Czech and English (and more generally Western) cultures. As such, it is not an insistence on traditional cultural values that would preserve Czech independence in the political future of the state, but rather a speedy incorporation of contemporaneity that would ensure a symbolic equality with the rest of Western Europe.

III. Martin Hilský: Shakespearean heteroglossia

The centrepiece of Hilský's public responses to both challenges is an attempt to dismantle the competitive edge of Hodek's and Sloupová's critique in denial of a definitive existence of one 'more true' or 'accurate' version of Shakespeare. Instead, Hilský freely admits – even endorses – a cultural climate where multiple interpretations and/or translations of the source text are viable at any given sociopolitical moment. His published remarks foreground a belief that translators should have different views of the original text, resulting in multiple, varied approaches to the actual linguistic transfer, leaving the distance between the source text and the target culture elastic and responsive. In this context, Hilský claims that both Sloupová and Hodek, in their insistence on a singular 'correct' translation of Shakespeare, miss the core of the translation process. As such, he argues, it would be difficult, if not impossible to claim that the difference between two translations is such that one could be understood as 'entirely good and the other entirely bad'. In fact, he adds:

> One correct translation of *Hamlet* does not and will not ever exist, in the same way that there is not and will not be an only 'correct' interpretation of this great and mysterious play. Every new translation of *Hamlet*, similarly to its every new interpretation, is the *seeking* of something that can never be fully discovered. Still, this process itself has a meaning, and I would argue that this non-translatability and non-interpretability of *Hamlet* is the catalyst of the never ending process of *discovery* of this ever-exciting text. (Hilský, 2000: 183)

In his own process of discovery as he translates Shakespeare texts, Hilský seeks to replicate the presumed distance between Shakespeare and Elizabethan theatre audiences, as well as to balance the assumed genius of the Shakespeare source text with the necessary immediate accessibility to contemporary Czech audiences. Inevitably, Hilský rejects Hodek's suggestion to translate Shakespeare's text into seventeenth-century Czech. Such a process would do 'the worst service to Shakespeare' as it would be 'deadening Shakespeare with deadened words'. By 'using the beautiful language of Comenius', he believes he would fail 'both Shakespeare and the Czech viewers. It would mean building an artificial wall between Shakespeare and the contemporary audience, a barrier that did not exist for Shakespeare's audiences'. To make the plays as accessible to contemporary Czech audiences as they once were to the Elizabethan and Jacobean Brits, to fulfil this 'central reason and necessity for modernization of Shakespeare', Hilský vows to translate the original text into 'contemporary speech' without sacrificing the dimensions of word play, double entendre and contemporary cultural allusion. It is evident, then, that rather than communicating to the audience the exact translation of words, which could result in an elitist exclusion of lay audiences from the enjoyment of Shakespeare's plays, Hilský aims to approximate historical consumption of Shakespeare's texts by erasing irrelevant archaisms and references which he deems unintelligible to contemporary audiences.

However, in the midst of this promoted relativity, Hilský's Shakespeare as a cultural icon remains largely untouched, lodged firmly as a centrepiece of universal artistic achievement. It is precisely Shakespeare's ardently defended interpretive flexibility that enables the multiplicity of productive engagements with the cultural capital inherent in the source material, endlessly adaptable to developing historical and social contexts, that forms the core of Shakespeare transcendence. Such relativization of Shakespeare interpretation, yoked to surrounding sociocultural contexts, parallels the multiplicity of the postcommunist society where, ideally, various proposals for viable postcommunist social organization engage in public democratic dialogue. Hilský's meticulously multiple Shakespeare, though not explicitly articulated as such, serves as a wishful embodiment of democratic subjectivity, wherein public consensus is not as important as open dialogue respectful of a multiplicity of viewpoints and experientially-delineated perspectives that, eventually, will yield a suitable template for postcommunist cultural organization.

The implicit link between multiplicity in Shakespeare and the postcommunist political process becomes more defined in Hilský's claim that it is only in the late twentieth and twenty-first centuries that translators have finally succeeded in penetrating the veil of Shakespeare's meanings and grew 'immensely closer to Shakespeare than were their predecessors in the nineteenth or the beginning of the twentieth centuries'. Challenging the view that it was the prominent patriotic Czech literary ancestors Hodek celebrates who understood 'true' Shakespeare, Hilský suggests that such closeness became possible only in the postcommunist moment. Once the Iron Curtain fell, Shakespeare's texts were freed from official socialist-realist interpretations and were subsequently exposed to the multiplicity of interpretive approaches that led to free mass popular consumption which, in tandem, enabled a wider perspective of humanity and its possibilities.

The engaged, faithful participation in the democratic political process, within Hilský's discourse, becomes equated with the faithful pursuance of the cultural meta-text, Shakespeare. As such, while denying the superiority of one translation of any other based on 'accuracy', leaving the overall translation perspective to the positionality of the translator, Hilský nevertheless delineates the parameters that will authorize translation work. In this case, the work of a 'faithful translator' becomes legitimate not through a particular linguistic approach, but through unwavering dedication to the source material, here Shakespeare. Indeed, translations, like his, that are 'created with a deep respect and love for Shakespeare' claim far more integrity and accuracy than any other approach to the cultural icon. To wit, Hilský claims that such 'loving' translation, though 'not immune to error', can 'go further away from the literal in order to capture the most accurate meaning possible, translation which simultaneously wishes to be contemporary and alive'. In other words, loyalty to Shakespeare, for all involved in the discussion, seems to provide the patina of certain authenticity of the translation in relation to the source text.

While no translation will capture all dimensions of Shakespeare's genius, Hilský argues, each can contribute valuable pieces to the overall debate. And so, playing on Hodek's language evoking inaccurate translation as national 'treason', Hilský argues 'we translators are all traitors ... but all of us betray in our own ways. It is important, I think, to betray learnedly and simultaneously with feeling, reverence, and love, to betray bravely and yet cautiously, to betray as beautifully as possible, but most importantly to betray truthfully'. Even as Hilský claims to evoke a relative and perhaps postmodern multiplicity of translations, he invokes a 'truth' in the translation process, a 'truth' that is as ambiguous and yet as central to the postcommunist political development as the 'truth' coined by Havel and celebrated by the masses of protesters in the fall of 1989, which was to make the centrepiece of postcommunist independent society.

IV. Czech Shakespeare translation and the postcommunist future

It is inconceivable that the challenges posed by Hodek and Sloupová to Hilský's work would not be provoked partially by the immense media attention that Hilský's work drew subsequent to the Jungmann Prize of 1999. Hilský's skyrocketing popularity among theatre professionals, who lined up for new Shakespeare translations for forthcoming productions, created parallel media interest, reflected in sustained reports of Hilský's scholarly (Shakespeare) endeavours, intensified programming of broadcast Shakespeare productions based on his work and periodic enquiries into his perspectives on Shakespeare. This constant attention, further fuelled by his well-received public proclamation of a goal to translate the complete works of Shakespeare by the 2016 quatercentenary of Shakespeare's death, transformed Hilský into a de facto Shakespeare spokesperson and provoked him to comment that he 'sometimes feel[s] like [he is] substituting for Shakespeare himself'.[5] Such public presence reinforced his image as a force in articulating the parameters of Czech engagement in Shakespeare and, in tandem, the parameters of post-communist culture.

The palpable dismay at the shape of Hilský's public presence at the break of the millennium evident in the arguments presented by his two challengers is further grounded in a spreading sense of futility of general intellectual endeavours. Within previous political frameworks of domination by a foreign empire, public intellectuals explicitly guarded the cultural core from imperial ideologies of the colonizer *du jour*. The widely championed perception of postcommunist freedom from foreign intervention in the Czech culture's future growth diminished this important role, forcefully separating the spheres of 'culture' and politics, and largely relegated humanist intellectuals to the towers of the academy or to the consumer-ruled sphere of book publication. This shift, as Robin Okey has observed, is intimately linked with the process of neoliberalization wherein this 'demotion of intellectuals' sense of importance and mission' is presented 'as outmoded baggage in a world of global expertise to which the region should aspire' (2004: 168). As a result, 'intellectual issues or debate no longer seem so central in communities where matters of cultural and national tradition have always enjoyed high prestige' and the intellectuals themselves, in submitting to the neoliberal relativization of their once-crucial input into the defining parameters of culture, 'risk being sidelined as mere "chattering classes" like their counterparts in the west' (ibid.: 169). Hilský's public presence, while undeniably prominent, has been limited to his Shakespeare expertise, not explicitly linked to the political future of the Czech society, attesting to the politicized separation of political and cultural spheres. One can only hope that, in a country traditionally reliant on its (predominantly humanist) intellectuals to weigh in on central political matters, Hilský's presence may provide a place mark for future generations of thinkers in the public sphere, if perhaps in different guises and conditions. The challenge to Hilský's work, while inevitably posing a challenge to the translator's work and public persona, simultaneously creates (if only negative) publicity for the intellectuals who partook of Hilský's success via their critique (or defence) of it.

Despite the voracious disagreements, the commitment to 'truth' as central to Czech postcommunist culture is common to all three participants of the 2001 translation debates. The variations of the precise configuration of this 'truth' are instructive in underscoring the ambiguous boundaries of the postcommunist political pathway. While Hodek finds this truth in the elitist inaccessibility crucial to the preservation of Czech traditional cultural values, Sloupová places it at the feet of the public sphere where all individuals should have access to its motivating version to participate meaningfully in a commonly-defined future. Eschewing a singular embodiment of postcommunist truth, Hilský embraces a dialectical multiplicity that is collectively to chart the Czech postcommunist future, trusting in the overall success of a democratic framework where competing ideas will not fail to eventually articulate a viable political course for the postcommunist nation.

That each of the participants seeks this truth via the cultural capital of Shakespeare underscores the cultural attempts to seek out firm models of human organization that could be productively informative in the otherwise bewildering postcommunist moment. Shakespeare's origins in the established West, extended by the assumed transcendental universality of his texts, are appropriated here to assist in the process of incorporating the postcommunist Czech Republic into the imaginary community of culturally developed countries in no need of further Western development. As such, the evocation of Shakespeare as a proper template for postcommunist existence seeks to counterbalance Western pressures to mould the Czech culture according to EU accession requirements. After all, if the postcommunist Czech culture claims an unalienable affinity with the paragon of Western humanism, Shakespeare, what improvement is there to make?

4
Katharina 'Humanized': Abusing the 'Shrew' on Prague Stages

Among the wealth of Shakespeare plays on Czech stages, *The Taming of the Shrew* has harnessed remarkable popular permanence. As of 2002, *Taming*[1] was the third most performed Shakespeare play since the end of the Second World War, following only *A Midsummer Night's Dream* and *Twelfth Night*, preceding even such usual suspects as *Hamlet* or *Romeo and Juliet*. Whereas other plays experience intermittent waves of interest and indifference, *Taming* seems to enjoy steady popularity: since 1945, 78 productions have been evenly spread so that there has not been a time in recent Czech history when a fan of the play, willing to drive one or two hours to visit a regional theatre, could not see at least one, if not two different productions. In the 2000–1 theatre season, there were no less than six *Taming*s country-wide, three competing in Prague alone, all based on the most recent translation of the play by Martin Hilský. Even in the context of its usual popularity with Czech audiences, such a spike in performance of one of Shakespeare's works widely categorized in the West as a risky 'problem' play[2] seems intriguing.

On the most elementary level, this chapter simply follows my curiosity about this intensification of Czech *Taming*s to delve into the cultural context that would have made this play – and the issues it offers up for reflection – seem relevant and noteworthy to contemporary Czech audiences. Since the play comparatively considers several courtships resulting in marriage, it seems natural to contextualize the 2001 productions in public discourses of love, marriage and relevant normative gender roles that collectively delineate successful nuclear families responsible for establishing normative parameters for the next generations of Czech citizens. In its rumination on the gendered nature of an ideal domestic relationship, the play enables a public reflection on gendered power within the private family sphere liable to a reconfiguration of state control. While divested of the ideologically charged communist insistence on gender equality, gender roles came under renewed state scrutiny early in the first decade of the twenty-first century. First, gender norms came under scrutiny as the government sought ways to respond to rising unemployment and a falling birthrate. Simultaneously, politicians needed to revisit normative boundaries of gender in tending to the EU's recommendation that the Czech Republic revise its gender policy. My analysis of the *Taming* productions zeroes in on evidence of surfacing tensions between the besieged state, which sought to find productive ways to satisfy EU requirements with relevant legislation, and the general populace that felt betrayed in the promise of postcommunist freedom from state interference in the private sphere by the reassertion of the state's control in matters as private as gender identity and gendered behaviour.

As in previous chapters, the locale of the discourse of postcommunist (neoliberal) subjectivity – Shakespeare productions – adds a crucial dimension to the central debate. Shakespeare's status as a timeless repository of humanity here assists in transcending the contemporary moment of sociopolitical tensions and provides a counterweight to the power of the neo-imperial West that seeks to enact substantial changes in the daily life of Czech citizens. The *Taming* text, made more ambiguous by the earlier anonymous play *A Taming of a Shrew* as well as by its framing prologue ('Induction'), is furthermore particularly suited to a variety of credible interpretations that lend symbolic weight to sometimes diametrically opposing views on normative gender roles. The central plot, wherein Petruchio apparently modifies the behaviour of his wife through the denial of sleep, food and clothing, undeniably prescribes a transformation of Katharina's fiery challenge of her environment to obedient alignment with acceptable Renaissance models of submissive femininity. In the absence of clear stage directions and of an epilogue that would complete the frame introduced by the Induction, however, the sincerity of Katharina's final pronouncements has remained open to debate. Where some interpreters see a woman truly transformed, others discover one who has learned to subvert the framework that seeks to control her to her own ends. Because the text provides little definitive evidence of where it itself stands on the issue, the play has been interpreted on the stage as a corrective to unruly female behaviour, as a chilling critique of dominant masculinity and patriarchal frameworks or, most frequently, as a comic exaggeration of common gender relations.

In the scholarly world, feminist critics have long pointed to the troubling aspects of the 'taming' within the play and, perhaps more crucially, to the audience's easy acceptance of Petruchio's treatment of Katharina. In her analysis of the play that zeroes in on the abusive elements of Petruchio's taming that – in the classical trio of sleep, food and clothing deprivation – parallels timeless torture practices aimed at making human subjects pliable to interrogation, Emily Detmer has argued that in order to perceive the play as a comedy, 'readers and viewers must work to see domestic violence from the point of view of an abuser – that is, they must minimize the violence, and, at the same time, justify its use' (1997: 274). In a parallel argument, similarly centred on the dissonant notes of Petruchio's conduct towards his wife, Shirley Garner has pointed out that an audience can find the plot entertaining only if it accepts 'the premise ... that a shrewish woman is less than human, even less than a woman, [and] so may be treated as an animal'. Evoking the necessary but frequently ignored interplay of a production with the contextual expectations of an audience, Garner underscores the cultural misogyny necessary for the play's easy digestion:

> Could the taming of a 'shrew' be considered the proper subject of farce in any but a misogynist culture? How would we feel about a play entitled *The Taming of the Jew* or *The Taming of the Black?* I think we would be embarrassed by anti-Semitism or racism in a way that many of us are not by misogyny. (Garner, 1988: 109–10)

In this context, the frequency of comedic productions of the play in a given cultural context can work as a barometer to indicate the current state of gender relations. To wit, Diana Henderson has usefully established that Western Europe and North America produce film versions of *Taming* in times of intense backlash against feminism, so that 'the clustering of filmed *Shrews* correlates with those decades when feminism has induced conservative responses and when the media are actively encouraging women to find their pleasures at home' (1997: 150).

The intersection of the three Prague productions of Shakespeare's *Taming* with the larger Czech sociocultural framework uncovers similar social tensions, particularly within the complex web of post-communist subjectivity formation, negotiating both public spheres of (inter)national discourse of normative identity and behaviour boundaries with a private sense of citizenship and traditional conceptions of self, agency and efficacy. The increased Czech fixation on the abusive, coercive and unequal relationship explored in these productions certainly parallels a spike in domestic violence in the culture at large. This spike is itself a product of unique postcommunist social, political and economic currents flowing from the neoliberal transitional process that has undermined traditional frameworks of charting individual subjectivity without offering appropriate alternative models. And yet arguing that these Shakespeare productions merely reflect an insecure machismo of Czech men eager to displace their anger on their partners would be a gross simplification of a complex situation grown out of disparate postcommunist political and social developments.

In the transitional process, the Czech culture has sought to integrate a legacy of four decades of communist-imposed gender equality that required women's equal participation in the workplace even as it stripped men of their traditional venues of self-realization within the capitalist concept of career-driven ladders. The ideal of the female socialist worker as an equal partner of the male even in the traditionally male-dominated work spheres permeated official communist ideology that sought to indoctrinate generations of Czech citizens after the communist electoral landslide of 1948. Somewhat softened by a later emphasis on childbearing as a central service to the socialist state, evoked as a response to sliding birthrates, this ideology was nevertheless vehement in its insistence on gender equality. This communist gender philosophy was buttressed by social frameworks that provided support of the traditionally labour-intensive caregiver sphere to facilitate female participation in the workforce, ranging from free educational opportunities to guaranteed employment, free healthcare, childcare, ample and guaranteed maternity leave, and support for the elderly. Despite these socialist provisions, equality was far from achieved: while women and men were nominally equal, statistics suggested that actual employment patterns tended to duplicate traditionally-conceived gender patterns.[3] Much as the communist state provided public services assisting mothers in integrating into the workforce, overall it failed in impregnating the private sphere of the family, where women continued to be responsible for the overwhelming proportion of household care and childcare, creating conditions of 'double shifts'. Moreover, for many, predominantly urban female citizens, dissent towards the communist ideological hegemony frequently took the shape of individual enactment of ultra-feminine self-fashioning as an opposition to the communist model of the union-suited factory worker or kolkhoz farmer. While habituating Czech women to substantive social support enabled workforce participation, the four decades of communism failed in instilling the ideology of equality. On the contrary, the discourse of gender equality, which was made one of the central pillars of communist social and economic planning, became politically suspect as an artificial ideology devised abroad to control the masses of individual domestic subjects.

After the fall of communism, gender relations were expected to return to their more 'natural' state unsullied by the interference of the public state, providing citizens with a range of new choices, most notably adding the previously unprecedented opportunity for married heterosexual women to stay at home with offspring. Yet the state's postcommunist neoliberal practices of privatizing the social sphere by shifting the material and temporal cost of family care onto the families themselves marked the first forceful postcommunist intervention of the state into the core of their citizens' private lives. The state indeed championed the gendered ideal of the single (masculine) earner household and cut support for public daycare as an increased incentive to privatize care for the family, hoping to solve increasing male unemployment by a mass exodus of women from the workforce. However, the rising employment instability and the deregulation of cost of even the most basic life necessities made women's incomes within the sphere of the family indispensable; in some cases, in the context of the wide restructuring of the economy, industry and service sector, women became the sole wage earners within a family. The diminishing social support for families and the demands of the workplace resulted in a precipitously falling birthrate, as young couples postponed childbearing until economic security of the household could be ensured and, in overwhelming numbers, postponed – or even dispensed altogether with – marriage. In addition, in the general restructuring of society, women found a greater flexibility in possible careers opened to them after the fall of communism, albeit with less job security and fewer benefits. The domestic sphere encountered a paradox: while postcommunist ideology celebrated liberated heteronormative families of male breadwinners, nurturing housewives and their happy offspring, postcommunist social policy destabilized the nuclear family so that all but the upper managerial class was set up for gendered failure. Such wide disempowerment resulted in a significant crisis of masculinity, already introduced in the Introduction and discussed further in Chapter 5; of relevance in this chapter are the rising rates of domestic abuse that manifested growing gendered tensions of unrealized patriarchal dominance.

In this context, the pressure of the EU towards greater gender equality – while understandable from without the immediate circumstances as a wishful corrective to the deteriorating conditions in the country – succeeded only in exacerbating the already widening popular sense of betrayal by the postcommunist transitional process. Far from living the rosy free lives promised by the postcommunist (neo) liberationist rhetoric, the majority of the Czech population struggled to meet the daily existential needs of themselves and their dependants. The neoliberal interventions of the state, in many cases implemented on the recommendations of Western experts who were to 'assist' with the transitional process, which eliminated the security nets that would have ensured greater liberty and pursuit of general happiness as well as privatizing many previously public services, belied the promise of state non-interference in private matters. The EU rhetoric of gender equality, made compulsory for the candidacy of the Czech Republic into the EU, seemed to intensify such state intervention in citizens' private lives; moreover, the experience of previous, widely subverted, communist ideology of gender equality made this new Western rhetoric inherently suspect. The lack of material support to buttress such rhetoric made it laughable.

And laugh the Czech population did, albeit in private. The compulsory nature of the EU rhetoric, which mandated that relevant legislation be enacted prior to Czech accession to the EU, made open protest unwise. Instead, even as Czech legislators dutifully discussed satisfying the EU requirements, the wider Czech population slid into the well-established ruts of double consciousness, established during long centuries of colonial and imperial domination. And so, while new laws mandated a ceasefire in workplace sexual harassment, for instance, and Czech employers dutifully followed with professional development seminars that were to assist their employees in revising their behavioural patterns, jokes about *'sexualni haraseni'*, a linguistic equivalent of a 'sexual madness' (presumably of the Western nations that brought it to the negotiating table) proliferated while actual harassment intensified behind the closed doors of the average workplace. Meanwhile, debates over *'pozitivní diskriminace'* – literally translated as 'positive discrimination', designed as an equivalent of affirmative action – unfolded in the daily press. In October 2001, the Czech Associated Press reported that Vladimír Špidla, the secretary for labour and social affairs who would a year later become the prime minister of the newly-elected government, established a new committee for 'equal opportunities for men and women' (čtk, 2001). The press dutifully noted the 'distaste of his government colleagues' who thought such a committee unnecessary but relented in recognition of the committee 'as a sure signal to the European Union'. Far from considering actual inequalities in the country, the committee was openly presented to the public as a measure necessary in order to meet EU requirements, 'helping the implementation of the usual trend in the EU, especially in Scandinavian countries'.

Like the cited (male) members of the cabinet, most newspaper readers reacted with to this measure with thinly-veiled contempt. Pavel Zídek, for instance, wrote to discredit the EU gender-equality efforts, arguing that 'in reality, women have special rights while it is men who are often discriminated against'. Revealing a measure of intense conservative essentialist views of gender roles, Zídek further opined:

> most [women] behave according to the stereotypes of feminine gender roles. These questions [about women's discrimination] will logically disappear when women begin behaving differently: when they stop establishing families or when they institute an alternative division of domestic labor ... Considering that marriage (partnership) is a voluntary institution, it seems that women

find the typical family model in which a woman does most of the housework and childcare satisfactory.

In laying the responsibility for gender inequality at the feet of Nature – or, minimally, women's inherent needs – Zídek turns the blame to the other pole of the gender binary, arguing that service to the state, such as mandatory military service, makes men into 'indentured servants' of the state.

Like sexual harassment or positive discrimination, new legislation aimed to counterbalance practices of domestic violence was met with general ambivalence at best. It was not until 2001, incidentally the year of the *Taming* resurgence, that, for the first time in the history of the Czech Republic, 'domestic violence' was loosely defined as consistent abusive behaviour and data on domestic abuse was collected and released to the public. In April, all major Czech newspapers announced that research on domestic abuse was underway, spearheaded by its criminalization in Western countries (especially the US) (Lucie Kolářová, 2001). Foreshadowing likely future legislation, the media warned that, though specific domestic violence legislation was yet to be drafted at that point, abusing one's partner was illegal and could be, in the severest cases, prosecuted under a number of general clauses, such as murder, rape or assault.

The subsequent results of the study, released in May, showed a staggeringly widespread violent enforcement of masculine dominance in the Czech cultural context. Within the existing legal framework, domestic violence was registered only in extreme cases. Defined as a prosecutable 'felony only once the victim [was] unable to work for seven days following the incident', domestic violence had to be documented in three separate official grievances by the victim. Even in such circumstances, the law stipulated that the prosecution of the crime could be stopped at any time at the request of the plaintiff, reflecting the widespread belief that domestic 'problems' were best resolved in the private domestic sphere rather than the public courts. This in practice resulted in systemic victim intimidation (sib, 2001). Not surprisingly, only an estimated tenth of all known cases of domestic abuse ever reached court, and the actual number of violent incidents (both as defined in its violent extremity and generally as physical or psychological dominance) was expected to far exceed the number of 'known' cases reported either by the victim or by a third party such as the assisting physician (Šverdík, 2001).

Even though 'domestic abuse' referred only to extreme cases of physical violence, data collected primarily in the capital showed that one Czech in six admitted to experiencing domestic violence first hand, either as a victim or an aggressor. Nearly two-thirds (61 per cent) admitted to witnessing abuse or having heard of it in their immediate context. Despite the large proportion of subjects who have experienced or have been in close proximity to domestic violence, a full 64 per cent of those asked relegated violent practices outside the perceived national mainstream to the category of presumably uneducated, poverty-stricken and otherwise violent persons. A significant proportion of the surveyed subjects further admitted non-interference: nearly half (41 per cent) were uncertain whether they would (or should) assist if they heard a neighbour screaming or weeping, while six per cent were sure they would ignore it. Nearly a third (29 per cent) thought that domestic violence was merely evidence of a negotiated domestic relationship that should be tolerated and solved within the family (Dolanský, 2001b). Comparative regional studies uncovered geographical differences, suggesting a far greater incidence of domestic violence in rural areas outside the nation's capital, coinciding with the unequal distribution of material social support and employment opportunities: in some areas nearly half of the population (46 per cent) claimed unfamiliarity with the term 'domestic violence' itself, and only 15 per cent claimed to know what to do if they witnessed it (avo and tof, 2001). The research further pointed out a nationwide dearth of resources for victims of domestic violence, a lack of shelters and ineffective legal recourse against aggressors. The police and medical staff were found to be complicit with systemic abuse, either ignoring incidents of known violence, issuing warnings that tended to create further backlash against the victims (lds, 2001) or even recommending that victims cease to 'provoke' the aggressors to prevent future assaults (Bělíková, 2000). In overwhelming numbers, the data suggested that domestic violence was thought to belong to the private sphere of the family that should remain taboo to public interference or even discourse (sib, 2001).

While the presented data on domestic violence undeniably evidences the pervasiveness of the prevalent form of practice in the postcommunist Czech Republic, the coincidence of the EU-sponsored research process with the pressure to implement gender equality measures likely contributed to the respondents' insistence on keeping gender relations private. As sinister as it seems, it is perfectly possible, indeed probable, that the EU recommendations themselves created a conservative backlash of widespread popular protectionism of the private sphere. Whether this backlash translated into an increased incidence of violence is likely to be open to endless debate; it is indisputable, however, that the inflexible insistence of the EU on dictating the boundaries of normative private relations at least provided an excuse to continue destructive patterns of interpersonal relations in the most intimate spheres of citizens' lives and created a common defensive rhetoric that justified its persistence. Moreover, there is no question that postcommunist neoliberal measures implemented at the request of the West exasperated – if not triggered – the paradox of the domestic sphere wherein individuals were systemically set up for failure regarding expected domestic gendered norms. This domestic paradox closely parallels the paradox inherent in neoliberal practices themselves, which are built on the rhetoric of equal opportunity while materially stripping away means of reasonable survival, setting up the working and increasingly the middle class for collective failure, inertia and alienation. One can only speculate that the spike in the physical manifestation of this disenchantment in CEE reveals a forceful translation of the culturally expected – but difficult to achieve – economic gendered dominance in the domestic sphere into its crude physical equivalent.

The rhetoric of national independence rooted in individual citizens' agency to define the parameters of their domestic relations worked to subvert perceived foreign interference in private domestic affairs. While remaining bound in relative restraint from open articulation in the public sphere, it nevertheless circulated widely within the culture, creating currents of resistance within the new neo-imperial framework. The three Prague productions of *The Taming of the Shrew*, in this context, provide a case in point. Neither the productions, the professionals who created them, the media that reviewed them nor the audiences that attended them would be caught in open articulation of condoning domestic violence. It must be noted that, conveniently, Katharina's treatment technically does not meet the postcommunist Czech criteria of domestic abuse: she is neither beaten nor incapacitated for a week from physical injuries, and she certainly does not press charges. Nevertheless, the thrust of the productions themselves, their reviews and their profound box-office success alone revealed a community that celebrated, condoned and made light of a manifest systematic dominance of an outspoken female character by her husband. More prominently, the collective interest in the public performance of a private settling of a domestic discord revealed fierce protectionism of the private right to 'work out' partnership differences in the domestic sphere, without the interference of the state. The forceful nature of the process exhibited in *Taming*, which results in the comparative disenfranchisement of Katharina as the domestic roles for each member of the couple are delineated, seemed a relatively minuscule price to pay for the autonomy the couple wins from the normative community that surrounds them. The popular example of the central couple in *Taming* provided an extreme example that resourceful individuals can and will do whatever is necessary to become productive citizens, if only left to their own devices. Shakespeare's stature as a universal repository of humanity reinforced the weight of such nationalistic subversion, lending its cultural capital as a counterweight to the artificial 'political correctness' pouring over the national boundaries.

Banking on Shakespeare's cultural capital, each of the three *Taming* productions participated in the contextual tensions surrounding postcommunist renegotiations of individual identity, efficacy and agency within the private and public spheres by foregrounding its own angle of interpretation of the play. Generally striving to preserve the fiction of Katharina's taming as a subject of comedy – though with various levels of self-conscious cultural meta-commentary – the interpretations include presenting the play as a disturbing masculinist dream of the economically and socially disempowered Sly (Komedie), a conservative narrative of essentially gendered love that overcomes socially-imposed neoliberal obstacles (CD 94) and a farcical consideration of Katharina's powerful cooptation of the explicitly patriarchal framework that seeks to bind her (Summer Shakespeare Festival). Despite this variety, after a consideration of the productions, interviews I conducted with their directors and cast, and the ideologically-streamlined press reviews, I come to argue that, most obviously, the productions collectively participate in fueling a nostalgic Shakespeare-buttressed fantasy of 'universal' gender roles, wherein *The Taming of the Shrew* is evoked as a model of a necessary domination of an unruly female by a rightfully re-empowered male, who regains his masculinity and reclaims his rightful dominant position within the patriarchal family. More importantly, however, these productions claim the transcendence of Shakespeare to argue for the autonomy of individual citizens to fashion their lives and relationships away from larger social policy and foreign intervention.

I. Dočekal's masculinist dream

The first production of Shakespeare's *The Taming of the Shrew* I consider, directed by Michal Dočekal, was performed in a small but well known alternative theatre, Komedie, dedicated to non-traditional, often postmodern interpretations of a mostly traditional repertory.[4] It is historically the first Czech production to take full advantage of the play's Induction and of an insertion of a recently-translated epilogue of *A Taming of a Shrew*. Taken together, the prefatory and concluding material provides a meta-framework to suggest that the play itself enacts a mere dream of a drunken commoner Sly who, in an alcoholic near-unconsciousness, becomes his own idealistic masculine projection. This projection, Petruchio, is all that Sly is not: powerful, intimidating, self-confident, eloquent, always self-possessed and ideally poised for the methodical 'taming' of the passionate Katharina, in the dream sequence implicitly substituted for the dominating wife of Sly's waking life.

Dočekal's modernized production reinforced this dream quality throughout the performance; Sly's dream is framed as a manifest response to the postcommunist process of neoliberalization which exacerbates Sly's existing insecurities by material and symbolic disempowerment. The destructive characteristics of this 'dream', in its temptation of ultimately inaccessible material access and cultural status, are further underscored by a creative recasting of the hunt with which the prefatory Induction begins. While Shakespeare presents us with a lord hunting in the woods who stumbles upon an inebriated peasant and decides to play a trick on him when he awakens, Dočekal's lord represents an ambiguous but supernaturally-powerful entity that hunts *for humans*; that the stakes of this game are deadly is evidenced by the casual handling of a severed human head by one of the lord's attendants. While the purpose of the overall game remains unclear, Sly as prey evidently presents an intriguing challenge of masculinity. Encircling his unconscious body on a hospital stretcher, the lord with his attendants magically translate him into Petruchio, an epitome of powerful masculinity, lush with physical and intellectual power, good looks, easy influence over others and material wealth. In contrast to Shakespeare's text, these paranormal agents do not disappear from the play, but periodically resurface in situations where Sly's working-class background falls short of a credible performance of the aristocratic Petruchio, to provide coaching through the action. Katharina's taming, in this production, arises from an instigation of the lord who coaches Sly on the proper dominance of his future wife.

The relativity of the ethical value of the taming spectacle is further underscored by a periodic disruption of the narrative flow of the play, wherein the actors performing Sly/Petruchio and Katharina appear at the lip of the stage, in front of an abruptly closed curtain, to directly provoke the audience into critical thought. Through incessant repetition, Sly's Shakespeare-prescribed spectator commentary 'A good matter surely./Comes there any more of it? ... 'Tis a very excellent piece of work' (I.i.242–53) becomes a refrain of a tertiary level of the play's reference. The actors, suddenly seemingly freed from the constraints of their roles to address the spectators intimately through the insistent repetition of the line, raise questions about the implications of the audience's presumed enjoyment of Katharina's 'taming'. The translation of Shakespeare's lines, '*pěkný kus, na můj vkus*' (literally 'nice piece, according to my taste') proposes that the enjoyment of gendered violence, whether in artistic representation or in everyday life, is a question of '*vkus*' or, literally, individual taste. Highlighting this 'taste' evokes the cultural popularity of violent entertainment – ranging from action films to video games and toys – and the subsequent blurring of boundaries between 'virtual' reality and the everyday, wherein the violence of one person against another – particularly when it comes to domestic abuse – is widely condoned and relegated to the sphere of entertainment.

Sly's occasional insecurity parallels Katharina's split behavioural patterns. Her exaggerated independence, projected through torn overalls, construction boots, short spiky hair and confrontational attitude, reminiscent of the negative stereotypes of communist construction workers (see Figure 4.1), periodically reveals an air of insecurity about her position in a world that does not seem to be willing to

Figure 4.1 Baptista and Katharina. *Zkrocení zlé ženy (The Taming of the Shrew)*, dir. Michal Dočekal, Divadlo Komedie.

accept her on her own terms. Decidedly isolated, Katharina is ignored by her mafia-like family, spurned by Bianca's wooers and overpowered by the unexpected Petruchio, who forces his attentions upon her. Her visible discomfort increases at her subsequent unlooked-for wedding, where, stripped of her baggy overalls in favour of a white wedding dress, she winces as Petruchio asserts his right to be 'master of what is mine own', calling her his 'goods', 'chattels', 'house', 'household stuff', 'field', 'barn', 'horse', 'ox', 'ass', simply his 'anything' (VI.ii.218–21). The general futility of her existence within the impending married life is accentuated by a process of visions that greet her upon her arrival to Petruchio's house: a drudge dragging groceries; an exasperated mother leading screaming children; and a bent housekeeper weighed down with pails and brooms. As her white dress greys and tears with the tumultuous experience of initial wifehood at Petruchio's dominant hands, Katharina's vitality fades and her initial outbursts against her husband's chillingly methodical taming give way to desperate begging for the basic human needs of food, sleep and clothing. By the crowning moment of Shakespeare's taming progression – Petruchio's refusal of a gown made for Katharina's use – Katharina exhibits evidence of a mental breakdown.

Katharina's breakthrough in marital relations comes en route to her sister's wedding, when she discovers the implicit bargain of domestic peace in exchange for her intellectual independence, evidenced in her obedient admission that the sun is the moon and the elderly suitor to her sister, Hortensio, a young virginal maiden. Her ready – and even relieved – submission to her husband signals a turn in the state of events by offering an apparently stable, predictable framework in which Katharina can survive. In the final scene, where she is reunited with the rest of the characters gathered at her sister Bianca's wedding, Katharina evokes the ideal of submissive and desirable femininity (Figure 4.2): in a long, low-cut black sparkling gown, high heels, long loose hair and with make-up on her otherwise composed face, she quietly but firmly delivers her chilling final monologue in its lengthy entirety, painstakingly outlining the parameters of general wifely submission to each inherently deserving husband. The visible awe

Figure 4.2 Petruchio and Katharina. *Zkrocení zlé ženy (The Taming of the Shrew)*, dir. Michal Dočekal, Divadlo Komedie.

elicited by Katharina's submission from the other male characters present on stage underscores the desirability of her transformation that proved Petruchio as the true 'master,/That teacheth tricks eleven and twenty long/To tame a shrew and charm her chattering tongue' (IV.ii.56–8). In their respective embodiment of conservative gender roles, both Katharina and Petruchio seem to have progressed from their relatively instable, socially disruptive origins to model citizenship recognized and celebrated by their communities. As such, the embodiment of traditional gender roles temptingly offers a corrective to their erstwhile failings of alcoholism, vagrancy, insubordination, depression, insecurity and self-destructive behaviour.

Yet the dream elements of the entire production throw productive wrenches into any easy interpretation of the play. Sly's subsequent awakening, wherein he implausibly proclaims to implement his newly-learned taming in his marital relationship, directly undermines the otherwise didactic message of Petruchio's taming. The erstwhile presence of the tempting, destruction-threatening lord who had set the violent process in motion and left the action to subsequent self-perpetuation of the destructive cycle of male dominance leaves the narrative with an unpleasant premonition of likely destruction inherent in Sly's subsequent taming attempts. Indeed, the dream Katharina's glossy lifelessness suggests that, even as an ideal, the model of femininity lacks the depth necessary to human subjectivity. All in all, the production seems to offer a compelling antidote to the cultural context of domestic violence that informs the immediate cultural circumstances of the audience.

However, as evident from the interviews I have conducted with the director and cast and, even more so, the press reviews of the play, Dočekal's production evidently enabled several mutually conflicted readings. Each could be encapsulated with a simplified interpretation of Katharina's visual transformation within the play from a resisting bride to a submissive wife. Since the production dutifully follows Shakespeare's text in that the space between Katharina's and Bianca's weddings, which bookmark the period of Katharina's 'taming', is a mere week, the audience is presented with the dilemma of deciding which of Katharina's hairstyles – her short unorthodox spikes or ultra-feminine long abundance – is part of her 'natural' identity. The other, by implication, is part of a mere costume, an artificial reactive identity constructed in response to repressive social frameworks.

If we recognize Katharina's earlier hairdo, the short spikes, as part of her 'real' identity, we can see her later strategic disguise with a normatively feminine long-haired wig as a part of the production's critical commentary on the requirements of artificial normative femininity for a successful integration into the postcommunist society. Such a reading can strategically take advantage of the production's dream-like framework, the periodic disruptions of the taming scenes by the actors' evocation of taste and the added non-verbal scenes of drudgery within marriage to equate Sly/Petruchio's taming, as Emily Detmer has done, with Stockholm syndrome. In this specific form of abuse, a victim, much like a hostage isolated from others and afraid for her/his safety, responds positively to kindness from the captor/abuser, and 'finds the key to survival [is] to actively develop strategies for staying alive ... [since] alternating coercive threats and kindness sets up situations where victims actively look for ways to please rather than upset their captors' (Detmer, 1997: 284, 287). Detmer demonstrates that, wishing to avoid further abuse and increase the instances of kindness, such a victim 'begins to subordinate her wishes, ideas, and actions in order to stave off the next outburst. It often develops into a cycle of violence that is used to maintain positions of dominance and submission' (ibid.: 276, n. 9). Katharina's final appearance, within this interpretive framework, evidences the chilling, personality-destroying consequences of Petruchio's successful dominance, which substitute an empty shell of oppressive social, male-focused expectations for a breathing human being. Put simply, within this reading, Shakespeare can be said to warn against the chilling realities of domestic abuse resulting in the loss of selfhood and complete isolation from others.

The preferred interpretation of most reviewers, however, was to think of the early assertive Katharina as the one in a 'harsh' disguise, hiding from a hostile world her fragile 'natural' feminine self, an essence which blossoms only in response to Petruchio's forceful 'love'. Within this reading, the extreme dimensions of Petruchio's taming are necessary to break down the barriers that prevent Katharina from participating fully in human society. Displacing Petruchio's early explicit interest in Katharina's dowry as the primary drive of his 'wooing' onto an assumed developing mutual marital love, this reading equates the force of Petruchio's dominance with the force of his assumed love for his new wife. Within this context, the dream qualities of the production function as an educational corrective to Sly's non-normative behaviour, beneficially enabled by the lord's generous assistance. Thanks to this transformative experience, Sly can reclaim his rightful dominant position in the society and, subsequently, facilitate the return of his wife to similarly harmonious submissive position.

Though all the reviews of Dočekal's *Taming* were aware of the problematic status of the play, pointing to it as one often targeted by 'feminists', none addressed the possible ways in which the production might reflect tensions of gender relations in the contextual Czech society. Instead, a large majority engaged in unquestioning celebration of Katharina's liberating transformation to ideal womanhood. For instance, Illona Francková, writing for *Zemské Noviny*, sees the process of taming as one awakening the initially 'unwomanly' Katharina to her essential femininity and to the appropriate position of the 'submissive but triumphant' wife who 'will be subservient to her husband in speech [and action], but not thought'. Superimposing an assumption of alternative (and perhaps subversive) thought on Katharina's submissive speech and behaviour, Francková's argument embodies the cultural belief that women's subversive power in the patriarchal system is equal to, if not greater than, the physically manifest patriarchal power that Petruchio holds over his wife. Martina Musilová similarly points to Katharina's initial lack of femininity, 'unusual for Czech circumstances', to argue that Petruchio does not 'tame' Katharina but rather 'educates her and with a fatherly mastery forces her to accept a livable and life-saving order' (2000: 53). On the most obvious level, such reviews condone conservatively delineated hierarchical gender relations – maintained by domestic violence, when necessary – as relevant models of essential 'livable' and 'life-saving order'.

More importantly, however, these reviews evidence the unshakable cultural belief in an inherent agency of dominated subjects in the face of external repression. As if drawing a parallel between Petruchio and the inflexible, paternalistic, transformation-demanding EU that does not question the degree of its influence over its new acquisition, these reviews celebrate Katharina's ingenuity in negotiating a rough terrain. After all, the parallel between the disenfranchised Katharina and the resource-strapped Czechs without any room for negotiation within their marital contract to the EU is not that far-fetched. One could indeed spend an idle afternoon delighting in drawing analogies between Katharina's utter dependence on her husband's willingness to feed her and the EU's agricultural subsidies that exclude CEE; her lack of access to her own clothing and the EU's industrial protectionism that undermined CEE businesses that provided affordable life sustenance for CEE citizens; 'Katharina's' lack of sleep could easily result from the unending barrage of shifting demands and unrealistic deadlines imposed by the West. In this context, the reviewers' insistence that, even within the parameters of repressive social frameworks, dominated subjects can find productive ways to survive within the private sphere of independent thought, undetectable within the public sphere, becomes positively uplifting. Katharina's transformation in this framework provides a recipe for balancing the scales of external domination and internal integrity of dominated subjects: by providing Petruchio with what he wants, a pliable trophy sex-object, Katharina gains the means of material survival while preserving a level of integrity of her internal subjectivity.

Perhaps because of the wider cultural implications of the insistence on Katharina's undamaged personal agency, few critics question the concept of taming as abuse or the assumption of women's subordination as a cultural fact; what they see on stage seems to be only a slightly exaggerated representation of the usual interaction between the sexes. One voice that pointed to the problematic transformation of Katharina on stage was Richard Erml, writing for *Reflex*. In 'Totally tamed Katharina', Erml questions her final appearance when she arrives 'conventionally beautiful as if from a magazine, and delivers the final monologue describing the advantages of submission with a completely mechanical voice. Is that how a tamed woman should look: beautiful and spiritless?'. Yet, as much as Erml's review seems to read the play as a critique of current gender roles, it simultaneously assumes that Katharina *should* be tamed. While protesting against the template of a tamed woman displayed for public consumption (the cover of a magazine), the review does not consider the process of taming itself. In other words, instead of commenting on the violence of Katharina and Petruchio's marital relationship, Erml critiques the production's portrayal of the transformed Katharina who, successfully tamed, should have been made more palatable to the audience. The play, Erml concludes, 'shows ... how two people can overcome the obstacles of the world and live an acceptable life together. It is enough to not suppress the personality of the other, but to play a game for the outsiders: in the times of Shakespeare the game was submission'. Perhaps the only critical voice undermining Petruchio's 'fatherly affection', which for so many outweighs his merciless dominance of Katharina, comes from Vladimír Mikulka, who questions the framing of the play as a dream as an excuse for the brutality of the interactions between the two principal characters. Not even the love-at-first-sight that Mikulka saw in Katharina can completely erase the kind of methods Petruchio uses to dominate or the fact that, at the end, Katharina is 'definitively broken', in her gown, heels and make-up, 'ironically contradicting the preceding claims about the little importance of clothes and outer beauty'.

The perspectives of those responsible for the production, the director, producer and cast, surprisingly reflect a similar duality of perception of the central characters' relationship. In published interviews, the representative voices of the production – the director and the producer – eschewed implicit charges of politically-involved production to insist on the play's non-ideological qualities. In an interview with Kateřina Kolářová for *MF Dnes*, Daria Ullrychová, the producer of the show, argued that the *Taming* production is 'only theatre', disconnected from political ideology; in fact, she argues, the cast sought to 'free themselves of ideology as much as possible'. Dočekal, in turn, limited his public remarks to a statement that, for him, *Taming* is about 'the problematic interaction between men and women, a theme which is always current'. In staggered interviews that I conducted with him in July 2001 and June 2002, Dočekal created more explicit distance between himself and the abusive elements of the play which he categorized as a 'sexist, anti-feminist play ... a comedy in which nothing is funny'. Refusing to stage the play as a comedy, he had conceptualized the production around the critical masculinist dream enabled by the commissioned new epilogue; only then had he found 'the only possible entrance point into the play' of perceiving the motif of the central taming plot as 'a play for the sleazebag Sly, who is imagining that he is someone he is not'. For Dočekal, the play became meaningful only when the taming became a form of social critique, exposing rather than glorifying an insecure, aggressive male ego.

Despite Dočekal's critical perspective, the principal members of the cast, whom I interviewed over the summer of 2001, built their performance on different views of their respective characters. David Matásek, who performed Sly-turned-Petruchio, saw his role within larger cultural connections. Like many of the celebrating reviewers, he found in the play a warning to Sly and all other underperforming males to correct their ways and reclaim their rightful position within the nuclear family and the postcommunist society at large. In this context of 'improving the world', Katharina's taming, though marginally excessive, was seen as a useful tool that enabled her journey towards her natural self and her appropriate position in society, which many women in the Czech society at large are regretfully abandoning. Pointing to the falling Czech birthrate – 'there are on average 1.1 children for two people in this culture, it is maddening' – Matásek argued for the necessity of the 'natural' procreative and familial role for women, for the greater good and survival of society at large. Remembering with aversion the non-traditional roles for women formerly supported by the socialist regime, such as the 'monstrous' large-machinery operators and construction workers – eerily shadowed by Katharina's early appearance – Matásek argued that no woman could honestly desire such employment. Rather, the evidence that women did fulfil such non-traditional roles was the result 'either of the lack of men to fill the positions or some ideological coercion which claimed that a woman is equal to man in terms of productivity, and that she can be yoked like everyone else. They forgot that there is one immensely important thing: it is called the family'. Yet, far more than resistance to remnants of communist ideologically-driven practices from which some entrapped and misled females might need rescuing by suitably forceful males, Matásek's support for Petruchio's *Taming* ultimately pivoted on a distaste for Western influences that presumably disturb the otherwise more natural process of the postcommunist transition. It is the artificial and ultimately empty Western rhetoric of gender equality, equated with 'militant' Western feminism channelled through women's magazines such as *Elle* or *Cosmopolitan*, that has assisted in a deterioration in the quality of individual lives of postcommunist women and men, as thus brainwashed women 'put on their little suits and managerial positions in large companies, and again forgot all about the family'. For Matásek, the transformation of Katharina seems to signify liberation from forced performance of state-sanctified gender norms in favour of a more natural role that satisfies the couple's private needs even as it fits them productively into the larger society.

Alena Štréblová, who performed Katharina, reported struggling with her role of a tamed woman which she found initially incongruous with her unshaken assumption of Katharina's love for Petruchio. While she recognized the currency of *Taming* in its exploration of gender roles, which are 'always, in every culture, unstable, and so their definitions have to be constantly re-negotiated', she equally rejected the possibility that Katharina might be lastingly damaged by Petruchio's behaviour or that her love for him might lessen. Like many of the reviewers and Matásek, she perceived Katharina as in need of forceful awakening from an artificial existence delineated by societal expectation to bring to life her true femininity. Her performance of Katharina's character was further layered by the dream framework of the play within which Štréblová, in the first step, played an actor at the disposal of the lord who staged Sly's dream-world. As such a player, Štréblová was to perform Katharina as an improvisation in response to whatever behaviour Sly-as-Petruchio would exhibit. Perceiving *both* the player and Katharina as 'outcasts ... women who do not even want to be women because they are afraid of their own femininity', she sees the two characters blending at the very end into one, both falling in love with Petruchio and his wooing ways. Having finally been found – and rescued – by a powerful man who not only validates their feminine sides, but rewards them, they realize that they need not be afraid of their own vulnerability and come to terms with their bodies and personalities.

Further identifying with her character, Štréblová spoke of her admiration for 'ideal relationships' of women who seem capable of managing both a successful career and the domestic sphere, in which they 'always have a clean house, and make their husbands breakfast, but this is not demeaning to either one of the partners. On the contrary, the more the women stoop, the more they gain in individuality ... and that is how they earn their right to be respected by their partners'. Translated to the context of the play, it is exactly the player-as-Katharina's self-sacrifice expressed in the final monologue that should, in Štréblová's eyes, gain Petruchio's respect. Yet, even in the world of the play this created problems, as Matásek in rehearsal reportedly refused to enact this respect in the last scene and occasionally slipped into victorious and boastful snickering. Admitting that the production did not show much of the self-sacrificing mutuality she feels between Katharina and Petruchio, she faulted the play and the textual cuts for not allowing its free expression.

Despite Dočekal's critical agenda and aspects of the production that offered a critical commentary on current gender relations within the wider postcommunist Czech culture, the reviewers and even the principal actors maintained their commitment to conceptions of essential, hierarchically conceived gender roles that should serve as building blocks of postcommunist transitionality. Turning a blind eye to the subtleties of the production, which – in the case of the actors – they themselves assisted in crafting, they instead relied on Shakespeare's cultural capital to buttress their support for autonomous private interpersonal relations divergent from the larger EU-spearheaded rhetoric as a base for a viable postcommunist future.[5]

II. Happy endings at CD 94

The second production of *Taming of the Shrew* I wish to consider here, directed by Michal Lang in another small and mostly independent theatre, CD 94, advertised itself as 'traditional'. This label immediately evoked the authority of Shakespearean cultural capital with the attendant insinuations of the longevity and universality of domestic gender struggles, as well as the production's seamless continuation of the well-established and culturally endorsed approach to both the play and the outcome of its plot. Whereas the other two productions I discuss provided minimal contextual information about the play, Lang's audience was offered an exhaustive background on Shakespeare, Elizabethan England, Renaissance literature and the play itself in hefty programme notes distributed at the door. The play is firmly linked to the popular Shakespearean canon via several quotations from *A Midsummer Night's Dream* and *The Tempest* that underscore the dream quality of the play. The production simultaneously sought links to universal themes by including excerpts from African poetry, on the inner covers of the programme, about a passionate marriage where the man (as opposed to the woman) does much of the domestic work, suggesting that the theme of gender inequality in a relationship is ever-present yet not universally gender-inscribed.

Read less generously, this reference also warns more xenophobic members of the audience that a lack of vigilance in maintaining established domestic hierarchies can and does result in an undesirable reversal of power that undermines traditional masculinity.

Moreover, the substantial narrative introduction to the production in the programme notes seeks to preempt the audience's possible condemnation of the forthcoming rough 'taming' action and to garner its support for its normative outcomes as culturally acceptable. The audience is to be guided by two reportedly representative Renaissance narratives, which presumably help elucidate Shakespeare's position on wife-taming. First, the introduction reproduces in full an anonymous short tale of a 'proud princess' reformed through a forced marriage to a beggar in a kingdom distant from her own. Initially humiliated and humbled, the princess is subsequently rewarded for her new-found industry and frugality by a revelation of her beggar-husband as the local ruler. Restored to her former class status and relevant power, the princess-come-pauper-come-queen becomes the model of womanhood and wifehood to the entire kingdom and its future generations. The second instructional narrative, similarly featuring a forceful reformation of a wayward woman by a well-meaning, self-sacrificing husband, is a graphic sixteenth century ballad, 'A Merry Jest of a Shrewd and Curst Wife Lapped in Morel's Skin, For Her Good Behavior'. This ballad, well known to students of the English Renaissance, tells a story of a woman resistant to her subservient domestic position whom her husband beats and sews into a salted horse's hide until she repents and consents to perform her domestic duties. Both reproduced texts are framed by the accompanying commentary as evidence that, in light of the 'frequent' accounts of wife-taming in Renaissance literature, 'it is good to keep in mind that, in comparison with other tamers of Shakespeare's time, Petruchio's approach is very mild'. While Katharina is undoubtedly spared the physical pain of salt in fresh wounds, the elision of the production notes over the psychological toll of Petruchio's actions on Katharina elevates psychological torture-like abuse (sleep-, food- and clothing-deprivation as well as public humiliation) over condemned physical acts, making Katharina's treatment a subject of instructive, progressive, forward-looking comedy. The notes further buttress Shakespeare's status as a progressive playwright by pointing out that Katharina's taming does not result in her complete silence as 'she has the last word', which the notes claim was unusual in times celebrating female silence.[6]

Unlike Dočekal's version at Komedie, Lang's production suggested that men can perform their 'rightful' dominant roles in a household if they are allowed sufficient material resources and attendant freedom from normative societal constraints. The play's prologue, within which Sly is temporarily elevated to aristocratic status and wealth, establishes a framework within which the spectacle of Katharina's taming begins a performance for Sly-as-aristocrat who turns Petruchio in the ensuing action. Sly's participation here is voluntary, a virtuoso impromptu performance of a spectator-turned-actor by the unexpected challenge of trying his hand at taming the tempting prey, Katharina. Lang's Katharina initially embodies a stereotype of an uncoupled heterosexual female existentially unsettled by the prospect of spinsterhood who projects her rage with her own perceived unworthiness onto her interpersonal relationships. Clearly attracted to Petruchio, Katharina is ready to believe his declaration to find her 'passing gentle … sweet as springtime flowers' (II.i.235–9) made to the soundtrack of soft instrumental music. Her affection grows despite occasional relapses to assertive speech and holds through her subsequent taming: though the production cuts Hortensio's descriptive recounting of Petruchio's abuse of Katharina at the wedding, Petruchio's brutality surfaces immediately in the post-ceremony scene as he chillingly claims his new wife as his property to be disposed of as he sees fit.

Katharina's ready compliance with her husband's demands that she change her behaviour before the taming proper commences only accentuates this brutality. Most visibly, this is exemplified in her relentless attempts to perform satisfactorily her gendered role of wife in an aristocratic household, as she attempts to mediate between the head of the household, her husband, and (presumably their) servants. While the audience is included in the servants' complicity in Petruchio's taming scenes, and thereby in the community that seeks to overpower Katharina's self, Katharina herself repeatedly attempts to insert herself between her seemingly enraged husband who physically assaults his attendants and his apparent victims, bodily defending, for instance, the scolded weeping tailor whose gown, made for Katharina, Petruchio had forcefully rejected. Her claim that she 'never saw a better fashioned gown,/More quaint, more pleasing and more commendable' (IV.iii.101–2) asserts less her narcissistic preference of fashion rather than her presumed power to mitigate her husband's censorship of the tailor's workmanship. Petruchio's rejections of Katharina's persistent efforts at performing her gendered marital role suggest that he is not looking for a partner in his household, but for dominance. The fact that all of the servants who assist with the 'taming' process are without exception male similarly suggests the production's aim for gendered dominance, resulting in Katharina's utter humiliation, rather than the professed goal of 'correcting' her behaviour towards expected Renaissance gender norms of an obedient wife. Accordingly, Katharina's taming elicits her increasing dependence on her husband; like Štréblová's Katharina at Komedie, this Katharina becomes an unquestioning extension of Petruchio's will as she agrees to his demand to recognize day as night and Hortensio as a bashful young maiden. In contrast to the *Taming* at Komedie, however, Lang's production offers these developments as unquestioningly positive, and undeniably entertaining, landmarks in Petruchio's progress towards mastery.

Katharina's final appearance, in the dress and cap she was formerly denied, underscores the extent of her transformation. Looking lovingly and seriously at Petruchio, she delivers her final monologue as a genuine reprieve to both the women on stage, Bianca and the Widow, and to the women in the audience. In a gesture of respect for her new-found identity, Petruchio kneels before her, only to be raised and kissed as they retire offstage, presumably towards private consummation of marital bliss. Since Lang's production does not borrow the epilogue from *A Taming of a Shrew*, the play closes with a double curtain call, wherein Sly returns to his initial non-aristocratic self, which Katharina, now unmasked as a 'mere' actor, finds equally appealing to the aristocratic Petruchio. The extended performance of the curtain call blurs the boundary between reality and fiction as Sly and the player-actor playing Katharina throw each other many loving but sad looks, communicating their disappointment with the inevitable end of their successful courtship.

The critical response to this production overwhelmingly celebrated the link between the taming and the play's eventual 'happy end', in which the characters allegedly enact the ideal of mutual married love, a link that clearly hinges on the perceived positive transformation of Katharina towards an appealing – and compliant – model of womanhood. Pavel Čírtek, writing for *Právo,* likened Katharina to a predator, 'a wild, unbound creature which needs to be tamed'. Celebrating the true identity she was able to access through her husband's forceful intervention in her behavioural patterns, he lauds her as 'a woman who in the beginning only intuits, but at the end knows how to gain her seemingly lost pride'. Jana Soprová, writing for *Večerník Praha,* uses a similarly celebratory tone to erase negative implications of Petruchio's taming by reassuring 'the feminists' who have 'long included the *Taming of the Shrew* on their "index"' that we as the audience at the end:

> have a happy-end feeling that the two really like each other. Not even the feminists should be too angry. Although Katharina submitted, she gave the surprised Petruchio such a lesson that he most likely will not ever torment the lovely creature again.

Zdeněk Tichý, noting in the title of his review that the taming of the feisty Špalková (the actor cast as Katharina) is 'worth it', marvels that *Taming* at CD 94 is one of the 'merriest performances' around, presenting to us the 'encounter of two strong individualities in which no one wins. Their sparkly joust suddenly transforms into happiness of a joint play which results in mutual respect and love'.

Even the few who find the violent aspects of Petruchio's behaviour 'chilling' (Urbanová) or even 'brutal' (Šrámková) explicitly undermine the possible reflection of the existing cultural context within the play's representation of domestic relations. Describing the taming scenes as 'playful' and 'unreal', thanks to the play-in-a-play frame, Urbanová believes that Shakespeare's fictional world 'does not set a precedent for actual life'. Šrámková, in turn, who evidently reads the play's action within the context of normative Czech marital relationships, chalks up Petruchio's brutality to Katharina's initial unwillingness to 'budge'; once Katharina, in her wisdom, recognizes the futility of their fights and learns to submit to Petruchio's dominant will, she discovers the 'better and more rational alternative of the relationship – love'.

Despite the vehement denials of the play's relevance to the immediate cultural context, the reviewers' collective tendency to comment on the action as if gossiping about their neighbours' turbulent marital relations is undeniable. In instructive parallels to the statistics on Czech domestic abuse that revealed a persistent commitment to keeping turbulent marital relations private, many of the reviewers articulate explanations of the experienced representations of marital violence that seek to justify general non-interference in the abusive relationship. Similarly to Šrámková's blame of Katharina's 'stubbornness' for Petruchio's outbursts, wherein violence can be readily eradicated via wise wifely obedience to their husbands, Jana Soprová minimizes the staged abuse by arguing that Petruchio's act of kneeling subsequent to Katharina's delivery of the submissive monologue signals his repentance for his earlier acts and symbolizes a pledge of future non-violence. As such, it participates in the common (wishful) perception of the victim's power over their abuser's actions, a perception that puts the responsibility for the abuser's violence with the behaviour of the victim; in this case, Soprová suggests, Katharina's eventual submission to her husband subsequent to the week-long 'taming' process has the power to eradicate future violence, without the need for interference from the rest of the immediate community or the more remote but better legally-equipped state.

By his own admission, Lang directed this production of *Taming* with an eye to representing a coming of age for both central characters. In an interview I conducted with him in August 2001, the director called the taming process a 'gate of sorts' through which both characters – but particularly Petruchio – enter adulthood:

> when a man tames a woman, they both grow up. Even when the man's maturation is not readily seen, it *is*, there, it is. That is why I chose to do the play. When [Katharina] presents the monologue, when she takes it seriously, and Petruchio kneels, he is meek. That is maturity. That is the point at which the little boy becomes a man.

In addressing the relationship of the two central characters, Lang cites the pervasive and timeless gender conflicts that surpass categorization as merely a 'women's' or 'men's' issue. Categorically rejecting the view that his production's interpretation of Shakespeare's play foregrounded a hierarchical relationship, Lang argues that Katharina and Petruchio strive for full marital partnership that, however, does not duplicate separate spheres of marital power, so that an 'equal woman' does not become 'masculinized'.

Lang's commitment to the 'separate but equal' perception of essentialized gender relations is explicitly rooted in postcommunist resistance to Western neoliberal influences. Critiquing the tendencies of Western consumerism, which provides the centrepiece of neoliberal society, to 'infantilize' individual citizen subjects to make them more pliable to 'business' strategies of continued consumption, Lang argues for the necessary social re-establishment of meaningful rites of passage that would mark the progress of individuals from childhood to maturity, marked by a decisive ability to resist neoliberal ideology. Shakespeare's text, rooted in the less corrupt past of the English Renaissance, reportedly provides a helpful corrective to the softened and misled postcommunist society in presenting essential gender archetypes that intuitively provide corrective models to infantilized members of the audience. In this case, Shakespeare is made to come explicitly to the rescue of the deteriorating Czech society, as an explicit antidote to the overwhelming influences of Western neoliberalism taking root in the most intimate recesses of the audience's private lives.

The principal cast members, in individual interviews, more or less confirmed adhering to Lang's vision in their performance of their respective characters. Petra Špalková, who played Katharina, cited the play's anti-feminism as a cheeky recipe for 'how to deal with feminism in the Czech Republic'. Echoing Štréblová at Komedie, Špalková argued that Katharina – though substantially transformed – 'wins' at the end of the play by submitting to Petruchio, who 'understands, kneeling in front of her', bringing emotional connection, peace and 'the happiest marriage of the three on the stage'. Such mutuality in marriage would not be possible, Špalková elaborated, without Katharina's eventual happy acceptance of her feminine essence, 'cultivated' in the process of her response to Petruchio's taming, an essence that she initially rejected, preferring instead to use her 'good, masculine, practical brain'.

Špalková's delivery of the final monologue exemplifies the most pronounced area of struggle with her role. While perceiving the litany of submissive wifely duties as excessive in the current cultural moment, an excess she readily chalks up to Katharina's 'blabbering' subsequent to her sleep and food deprivation, she simultaneously used its critical elements to address masculine members of her audience. Rather then exhorting women to greater obedience, then,

Špalková's Katharina used the monologue's articulated assumptions of masculine protectionism (a true man is 'one that cares for thee/ And for thy maintenance commits his body/To painful labor both by sea and land,/To watch the night in storms, the day in cold' [V.ii.152–4]) to point out the shortcomings of the infantilized postcommunist managerial class of 'men in the audience with their briefcases, cell phones, and manicured hands'. Widening the net of social critique, Špalková's Katharina implies that Czech women's presumably unfeminine independent behaviour arises from the softening of Westernized Czech masculinity, which has substituted bodily comforts for true manly pursuits. Like an overwhelming number of reviewers and theatre professionals, then, Špalková centres the viable future of the Czech postcommunist society on a wide performativity of essential, dominant masculinity that would enact collective resistance to the anomie of increasingly pervasive postcommunist, West-elicited neoliberal subjectivity.

The conglomerate of responses to the Komedie and CD 94 productions suggests that, while not a centrepiece to the re-empowerment of Czech men, domestic violence comes to be perceived as a necessary evil that utilizes a perhaps questionable means to a worthwhile end. This end of stable, well-ordered and well-connected nuclear families here provides some hope that such a re-centring of nuclear power in the traditional locus of the family and society would successfully undermine the perceived 'divide-and-conquer' approach of Western neoliberalism.

III. A 'seedy farce' at the Summer Shakespeare Festival

The third and last *Taming* that I wish to address in this chapter, Vladimír Strnisko's *Taming* at the open-air Prague Summer Shakespeare Festival (SSF), provides an intriguing contrast to the productions at Komedie and CD 94, in that it works to reverse the central gendered power dynamic to undermine the results of Katharina's 'taming'. Taking advantage of the summer availability of actors from a variety of theatres, Strnisko created a multinational cast of actors, where actors from the Slovak National Theatre played the Minola family and Czech actors the rest of the roles.[7] Without the framework of the prologue and epilogue, Strnisko's *Taming* inevitably struggles with an acceptable version of the taming story not cushioned by the illusion of a dream. Whereas both performances of *Taming* at Komedie and CD 94 provided chilling interludes as an interruption of the presumed hilarity of jokes, the SSF performance was lively and fast-paced, entertaining the audience with unending stunts which rarely paused for a moment of reflection. And yet, instead of dishing out a straightforward comedy, the production successfully undermined some of the central gendered assumptions usually associated with the play and subsequently stirred the wrath of the reviewing press.

Instead of the usual cast of youthful actors in the two principal roles, where Petruchio's slight advantage in years seems to confirm his claim to superiority and control over Katharina, Strnisko's Katharina was acted by Anna Javorková, a well-known and experienced mature actress with an impressive resumé of lead theatre and film roles. Cast against Michal Dlouhý as a youthful Petruchio, a veritable teenage heart-throb famous at that time predominantly for his portrayal of handsome princes in films targeting young audiences, Javorková's Katharina easily dominated the power-dimensions of the stage with her self-possessed, confident and experienced manner. Her costume of red leather hose under a laced airy light-coloured dress, which accentuated her as the most colourful character on stage, further contrasted with Petruchio's scruffy subdued blues.

In stark challenge to the established expectations of the standard self-possessed, dominant tamer embodied in both the Komedie and CD 94 performances, Petruchio's verbal assertions of dominant masculinity in the SSF performance were systematically destroyed by his obvious inexperience, awkwardness and the non-verbal actions of other characters on stage. Whereas the other two Petruchios exerted considerable influence over their surroundings, exemplified by their easy dominance of their households and servants, this Petruchio fails to control even his body servant Grumio. Perpetually inebriated and hopelessly insolent, Grumio consistently dresses down the masculine claims of his master. Early in the play, as Petruchio seeks to impress Bianca's suitors by his own masculine boasts, Grumio hurls a series of discrediting remarks, grunts and headshakes that punctuate Petruchio's lofty speech:

> Have I not in my time heard lions roar?
> Have I not heard the sea, puffed up with winds,

Rage like an angry boar chafèd with sweat?
Have I not heard great ordnance in the field,
And heaven's artillery thunder in the skies?
Have I not in a pitchèd battle heard
Loud 'larums, neighing steeds, and trumpet's clang? (I.ii.189–201)

Since he is unable to control his immediate attendants, it is not surprising that Petruchio loses his initial bravado early in the wooing scenes, even prior to his first encounter with Katharina, as he contemplates the physical marks Katharina left on Hortensio as a reward for his efforts to teach her lute fingering. Petruchio's attempt to sneak offstage and depart before his first trial of masculinity is foiled by Katharina's father Baptista who, desperate to rid his household of Katharina's dominance, literally drags Petruchio to his first encounter with his future bride. With the exception of few moments in the taming sequence, Katharina easily dominates the action throughout the play. Forward in both speech and behaviour, she slowly but surely stalks Petruchio in the wooing scenes with the ease of a jesting panther, bristling with wit as she readily responds to his clumsily-delivered lines. On a perpetual defensive, Petruchio struggles both for pertinent comebacks and for suitable hiding places, steadying his tremours on stage props. In this wooing scene, analogous to a lioness's amused game with a condemned rabbit, Katharina's presence is utterly incapacitating; incapable of coherent speech as long as she is on stage, Petruchio finds words only when, seemingly bored, Katharina exits.

While the other two Petruchios assume full control of their Katharinas with the transfer of legal guardianship at the moment of marriage and proceed towards forceful taming in Katharina's new household, the SSF Petruchio remains incapable of asserting his expected dominance. At his wedding, he props up his courage with alcohol, which lends his outbursts an aura of desperation and insecurity. The wedding guests' reaction to his raving, as they attempt to reclaim Katharina from his grasp, reinforces the notion of the alarming nature of his violence, incompatible with the values their community embodies. The subsequent taming scenes in Petruchio's manor equally undermine the usual brutality of Petruchio's control of Katharina as they prove ineffective. Fast-paced, eliciting squeals of laughter from the audience as Petruchio attempts to devise ways to disrupt Katharina's dominance in the relationship, the taming scenes reveal Petruchio's lack of experience in dominating a wife: breaking the fourth wall, Petruchio even desperately solicits the audience's help: 'He that knows better how to tame a shrew,/Now let him speak' (IV.ii.179–80). The effects of Petruchio's actions, no matter how ingenious in their approximation of well-tried mechanisms of torture, were minimal. Though indisputably hungry, Katharina seemed to be little affected beyond becoming marginally cranky. Instead of being 'taught' how to fit into the world around her, a feature of the play many critics have argued is redemptive of the taming as it 'enables' the character to 'begin functioning' in her surroundings, Strnisko's Katharina encountered in Petruchio's taming only social maxims that she had long rejected. With exasperation enviable by every teenager bored by their parents' repeated exhortations, Katharina petulantly interrupted Petruchio's moralizing to finish his recital of maxims that should delineate her subjectivity (IV.iv.164–8). Whereas the sun/moon scene traditionally marks the final breakdown of Katharina's subjectivity, Strnisko's production transformed this climax into a congenial moment of collective inebriation. As an affectionate mother temporarily satisfying a silly child, Katharina gleefully described the beauties of the sun-become-moon and took even more apparent pleasure in confusing Hortensio with her suggestive praise of his youthful maiden qualities.

The futility of Petruchio's 'taming' became most visible at Bianca's wedding where, traditionally, all Katharinas perform their transformed obedient femininity. Having shed the long white wedding overcoat that had hindered her movements throughout the taming scenes, Katharina reappears in her former leather costume. Though openly affectionate with her pet Petruchio, she remains her initial assertive self with everyone else. When it comes to the ultimate test of wifely obedience, Petruchio exhibits as much uncertainty about her expected response as he had betrayed in the first wooing scene. Her obedient appearance when called leaves him more surprised – and open-mouthed – than any of the other characters on stage. Her subsequent compliance with his requests seem equally subject to her own whimsical – and entirely unexpected – decision to toy with her husband. Katharina's playful subservience in this production does not extend to Shakespeare's chilling final monologue, which she cuts drastically to run off with her disbelieving sister to chat in the wings, presumably about the gullibility of their husbands. Petruchio's masculine triumph in front of the other characters on stage, so unexpectedly gained, immediately dims as explosions of female laughter sound from the wings where Katharina had set up an amiable female circle. The stage lights dim on this production as Petruchio's initial masculine uncertainty spreads to all the other men, suddenly shaken in their sense of dominance in their own domestic relationships.

Of the three performances, this one fared the least well with critics. While the *Tamings* at Komedie and CD 94 were celebrated by the press, the only praise reviewers could find for Strnisko's version regarded the comic aspects of the production. Rather than as a serious production of Shakespeare, Strnisko's take on the play was perceived as merely light entertainment for otherwise empty summer evenings, providing good laughs for bored audiences not willing to engage with serious matters. In a perhaps stunning turn of events, many of the reviewers, without providing further detail, focused on the production's perceived degradation of Shakespeare's source text. Despite the production's use of the identical translation of Shakespeare's *Taming* by Martin Hilský used by the productions at Komedie and CD 94, reviewers argued that the language of the script was too vulgar and too sexual to represent faithfully the core values of Shakespeare's play. 'True' Shakespeare, this criticism insinuated, is less earthy, less focused on sex and, most importantly, does not give room for Katharina's subversive actions. The majority of media reviewers commented that the production degraded 'one of the most beautiful of Shakespeare's plays' (Vlastník, 2000) to a 'seedy farce' (Machalická, 2000; Jeníková, 2000b). This involved and highly charged response suggests that, for these critics at least, the idealized Shakespeare has been tainted, departing as the production does from the traditional venues of interpretation.

In an interview, Strnisko readily allowed that he aimed at a rowdier '*comedia dell arte*' production in the open air than one he would attempt on a permanent stage, where, he was quick to assure me, 'it would be much, much more serious'. The production was based on his grounded belief that, whatever the circulating rhetoric of proper gender relations, it is women who are in control of their relationships and, by extension, of society at large. Men, like his production's Petruchio, merely follow along while seeking to hold on to appearances of control; Petruchio, in his view, 'talks about [Katharina's successful taming] as a done deal, though a minute ago he still doubted it. He happens to do one thing right and immediately pretends that he has been working on it for a long, long time. But he knows, deep in his soul, that he has Katharina to thank for it'. As to the vulgarity cited by the hostile reviewers, Strnisko readily pointed out its presence in Shakespeare's source text and noted, with a quiet chuckle, that none of the actual audiences in Slovakia, Moravia or Prague, where the production was staged, 'seemed to mind'.

Javorková, incidentally Strnisko's wife offstage, echoed the director's view of gender relations in the wider society. 'When a woman is smart', she offered, 'she will manipulate the man even though he thinks he has her tamed. The opposite is true'. Perceiving the taming sequences as evidence of a breakdown of effective communication between the two partners, a breakdown that Petruchio in his frustration unsuccessfully seeks to right with violence, she argues that the violent scenes can never work as they are written to produce a mutually fulfilling marital relationship, but need to be radically revised. Pointing to Katharina's radical revision of the text of the infamous final monologue detailing wifely duties, a revision that in this production deleted the majority of Shakespeare's lines and undermined the rest by Katharina's playful exaggeration of reciting tired maxims of hierarchical gender relations, Javorková underscored the necessity of balancing each partner's contribution to the marital sphere for a successful marriage rather than adherence to preconceived social models of love and marriage.

An interview with the dashing Michal Dlouhý, who performed Petruchio both at CD 94 and the SSF, revealed an intriguing individual approach to the role that transcended the directorial perspective of both Lang and Strnisko. Conceding that women unfortunately do run the Czech society, Dlouhý speculated – in a witty topical reference to Javorková and Strnisko's marriage offstage – that female dominance may be responsible for *Taming*'s popularity in the culture, as female actors eager to play Katharina successfully manipulate directors and producers who subsequently give way. In this context, it is the role of the besieged man to defend his integrity and limited social power. Each of his Petruchios, to varied extents, mirrors this belief. In the SSF production, which is more explicit about the reality that 'the women always tame the man', Petruchio claimed victory through his successful reactive response to Katharina's dominant assault as well as his successful bravado in front of the other male characters: 'Petruchio won. He won for himself. He got what he wanted. And he won in front of the others.' Contrary to the perception of Petruchio's inexperience in Strnisko's production, Dlouhý maintains that his Petruchio was not nervous or afraid of Katharina, but was instead fully in control of his – and by extension of her – actions. Queried about Petruchio's apparent discomfort around her on stage, Dlouhý proposed that Petruchio was perhaps simulating insecurity to learn 'her true colors so that he knows exactly how to tame her'. He argued that the fully tamed Katharina is likely grateful for the change, since 'all women want to be tamed' and 'returned to the world' of essential and hierarchical gender relations.

Yet, comparing the two productions, Dlouhý favoured Lang's interpretation of Shakespeare's play that provided a fictional fantasy of overcoming women's dominance, perhaps as a suitable antidote to women's general rule in society. There, Dlouhý argued, Petruchio:

> is not mean, he is not brutal. What you see is self-confidence. All the taming, all the things he does to Katharina he does only for her, to change her view of the world, to make her approachable, to make her human. He is not there to break her; he breaks her only to the extent at which she changes into a normal, pleasant person who is able to take in her surroundings.

Overall, Dlouhý lauds the perceived corrective thrust of Shakespeare's *Taming*, which could assist in suppressing 'a bit women's emancipation, since it is not natural'. Evoking the postcommunist context which has seen a more pronounced diversification of gendered behaviours, Dlouhý suggests that Shakespeare 'shows that women do not have to do all that men do … it has been a trend in the last few years that women began meddling in gentlemen's matters. This play could be a healthy mirror showing that a woman is a woman, a man is a man, and mixing "certain things" is not appropriate'.

IV. Towards a gendered postcommunist society

In its apparent incongruence with the actual performance of the two Petruchios, Dlouhý's commentary valuably exemplifies prevalent attitudes towards gender surfacing in the transitional postcommunist

Czech culture. Like the majority of theatre professionals and the press reviews of the three Prague *Tamings*, Dlouhý rearticulated Shakespeare's text to fit the purposes of the wider contexts of postcommunist social organization. Eschewing evidence that would threaten Petruchio's masculine dominance, whether in explicit evidence of its failings or negative implications of violent masculine action, Dlouhý – like Matásek, Štréblová, Špalková, Lang and the overwhelming majority of reviewers – rearticulated the *Taming* productions to contribute to the wider popular resistance to postcommunist socio-economic developments that have overwhelmingly disempowered large proportions of the population and have been particularly felt in the disempowerment of large numbers of men unable to exercise their expected agency of family providers and supporters. In this process of ideological re-enfranchisement of the male population, the domination of women represents a symbolic substitute to the control of wider postcommunist circumstances. The evocation of Shakespearean cultural capital further lends valuable (though, as the voracious critique of the SSF production suggests, precarious) ideological currency to this subversive cultural countermove to the EU's attempts to influence normative boundaries of gendered identity and its manifestation in the private relationships of Czech citizens.

While the majority (64 per cent) of the respondents queried about Czech domestic abuse in 2001 displaced extreme violence from mainstream society on the uneducated, poverty-stricken working class, thus distancing the core of Czech national identity from violent manipulations of power inequalities within the private domestic sphere, the overwhelming celebration of the chillingly brutal 'taming' of Katharina at Komedie and CD 94 reveals a far more pervasive commitment to securing the means of ensuring a social organization that does not unquestioningly implement the normative frameworks dictated by the EU. Indeed, the autonomy of social organization from external influences is key in all three productions, paradoxically most palpable in Strnisko's SSF version that questions the viability of traditional gender roles to argue for autonomy of a non-traditional relationship between Katharina and Petruchio. Despite their interpretive differences – which include such disparate interpretations as Dočekal's pro-feminist stance critical of interpersonal abuse, Lang's insistence on the necessity of violent interpersonal feuding for productive adulthood, and Strnisko's bid for relative autonomy of various domestic arrangements – these *Taming* productions cumulatively underscore the importance of interpersonal negotiation of the domestic sphere without external interference.

That these issues surface so powerfully in productions of Shakespeare is no coincidence. The cultural capital of the Avon Bard serves as a convenient distancing tool that sheathes the productions in the aura of timelessness and universality that divorces them from the postcommunist political sphere. If Amelia Howe Kritzer is correct in defining political theatre as theatre that addresses issues visible in the public political sphere (2008: 9), the three Shakespeare *Taming*s addressed here would undeniably fall into the category of political art. Yet the mere evocation of Shakespeare's postcommunist aesthetic autonomy, so voraciously defended by both theatre professionals and Shakespeare scholars, helps shield both the productions and their subject matter – proper gender roles and acceptable force to maintain them – from public political scrutiny. In itself, this distance serves as a reminder of the wishful separation of art from politics in the allegedly ideology-free postcommunist era. More importantly, it signals cultural determination to move private issues – such as models of gendered political subjectivity and interpersonal interactions – from the floor of the Czech senate and the EU Parliament back to the domestic sphere.

5

Politics of Desire: Postcommunist Czech Shakespeare and Non-normative Masculinity

If hierarchical essential gender relations, discussed in the previous chapter, have become a crucial component of postcommunist subjectivity resistant to external manipulation by normative requirements of the EU, heteronormativity has become its undeniable centrepiece. The precariously ambivalent postcommunist cultural framework discourages non-normative gender relations within heterosexual relationships. It is similarly ambivalent about non-normative variants of sexuality, particularly those that challenge dominant heteronormative masculinity, which has traditionally been seen as being at the core of national citizenship. This ambivalence is simultaneously proactive, in that it reflects cultural attempts at defining the boundaries of the postcommunist Czech society, and reactive in its persistent resistance to external pressures on the process of postcommunist development. In the context of guiding the restructuring of normative boundaries of gendered subjectivity in its candidate states, the EU has required legislation granting equal citizenship and social rights to homosexually-identified citizens as a condition of the eventual accession of CEE countries to the Union. In a parallel responsive move, Czech society – fuelled by a neoliberally-provoked crisis of masculinity – performed the double-move of seemingly accepting the forthcoming relevant legislation while intensifying its normative masculine frameworks of cultural reference.

This chapter builds on the Montmartre *Sonnets* anecdote with which this volume begins – wherein the Czech public and media effectively boycotted explicit homoeroticism in a production of Shakespeare's *Sonnets* – to explore further the implications of the performativity of non-normative masculine subjectivity in the postcommunist Czech context, with a particular emphasis on male homoeroticism and homosexuality. First, I will contextualize the present discourse of sexuality within a brief outline of the discourse of masculine sexuality as part-and-parcel of the historical discourse of Czech nation-building. I will thereafter consider the sociopolitical and cultural dimensions of performing (non-)normative masculinities (and sexualities) in postcommunist Czech theatre, both popular and Shakespeare. My analysis of non-normative masculinities in Shakespeare begins with a consideration of a rare 1997 production of *The Merchant of Venice* which transposed the central conflict between the Venetians (spiced up with a male-to-female transsexual Portia) and Shylock to a contemporary postcommunist setting where a morally-ambiguous drug-trafficking mafia challenges the orthodoxy and fiscal conservatism of the normative social status quo. Next, I link the radical *Merchant* production with the non-normative dimensions of the aforementioned Montmartre *Sonnets* in a short survey of traditional *Sonnet* performances that collectively reinforce dominant heteronormative models of masculinity. Finally, I abandon the survey of non-normative masculine sexuality in postcommunist Shakespeare in favour of a detailed case study of two representative postcommunist productions of *Twelfth Night*. Rich in variants of gendered sexuality, the play in postcommunist Czech performance enables a complex foray into negotiating the boundaries of postcommunist normative subjectivity in the highly charged cultural space of contemporary Czech theatre.

Eventually, I come to argue that the normative shapes postcommunist theatrical representations of masculinity take are both dependent on the overall sociopolitical context (an argument made throughout this book) and also on the cultural space where non-normative identity is portrayed. While the general audience seems ready to engage with representations of non-normative sexuality as an exaggerated sensation, especially on the level of the popular culture of the musical, it generally refuses to condone positive portrayals of non-normative masculinity and masculine sexuality when presented as an inherent part of the venerated canon of literary works that are to speak to the essence of humanity. A male-to-female transsexual, especially when surrounded by a drug culture of irresponsible, excessive Venetians, is clearly a creative addition to Shakespeare's imagination in

The Merchant of Venice that potentially uses homoeroticism as but one of many negative attributes of the morally-debauched mafia. As such, it can either be discarded as a 'creative whim' of an eccentric director or explained as a didactic warning against the dangers of a hasty implementation of neoliberal practices that undermine cultural values crucial to a more-or-less organized society. Similarly, within the multiplicity of changes enacted among the constantly developing identities of the *Twelfth Night* characters, the portrayal of non-normative masculinities can serve as a part of a rites-of-passage process where individual characters mature towards organized heteronormative relationships. In contrast, in the Montmartre production of *The Sonnets* where the only characters are the two male lovers, no such displacement is feasible; instead, the audience is faced with a personal and pressing decision either to accept homosexuality as part-and-parcel of human nature or to deny the existence of the issue altogether. As such, rather than exploring the effect of considering multiple iterations of masculine identity, most Shakespeare productions seek to reinforce the traditional powerful heteronormative masculine code that, when projected into the sphere of the larger postcommunist culture, can serve as a suitable template for independent and powerful postcommunist citizenship.

The redefinition of postcommunist Czech citizenship – and masculinity at its centre – straddles an ideological conflict of differing perceptions of the West. One perception is upheld as essential Westernism that can serve as a future cultural model; the other is perceived as the existing cultural and political models of the established 'Western' nations, perhaps originally rooted in the essential Westernism but ultimately corrupted by the process of cultural and political developments. While the Czech Republic has intensively lobbied for admittance to the Western EU, it has simultaneously sought to establish an individual, well-delineated national identity that can be seen as separate (and preferably superior to) existing Western cultural models. Much popular cultural smugness has rested on resisting the rhetoric of 'political correctness' of the established West, particularly its multiculturalist, feminist and other dimensions of widening perceptions of national belonging and agency. Such resistance to hybridity is not unusual in countries recovering from long-standing state, cultural, political and ideological domination by (an)other national entity, as Edward Said points out in the introduction to *Culture and*

Imperialism. In arguing that nation and culture become, in such tense historical moments, often as enmeshed as to seem indistinguishable, (re)building a nation on paths that are to 'return' it to its 'culture and tradition', Said claims that 'these "returns" accompany rigorous codes of intellectual and moral behavior that are opposed to the permissiveness associated with such relatively liberal philosophies as multiculturalism and hybridity' (1994: xiii).

In large part, the anxiety about homosexuality is fuelled by a perceived crisis of masculinity and 'good' citizenship which is blamed for an increased prevalence of unconventional gender relations that have in turn been identified as grave threats to orderly society. Paradoxically, as many critics of neoliberalism have emphatically argued,[1] the displacement of blame for deteriorating living conditions from economic to social factors, from neoliberalism to liberalism, is a crucial practice of neoliberal self-preservation. Purposefully conflating private economic freedom (to consume and profit) with civil liberties (particularly in terms of race, gender and sexuality), neoconservative neoliberals tend to take advantage of the social crises facilitated by neoliberal practices to implement conservative movements toward restoring traditional normative boundaries in the private sphere. As implementation of neoliberal practices of nominally 'free trade' begins producing astronomical profits for a small proportion of private investors, while the disenfranchised rest of the population begins to feel its own economic decline as well as the decline of the public sphere, conservative social, political and economic agents promptly allocate blame to the non-normative forces in a society while continuing to champion core neoliberal practices. This trend is not specific to CEE; one only has to remember a few examples from the USA, for instance, the intense backlash against immigrant communities at times of economic uncertainty in the USA or the redoubled efforts to reinstitute heteronormative marriage as a viable strategy for reversing the deterioration of the middle class. In the circumstances of the Czech state, and CEE in general, where non-negotiability of the conditions of neoliberal structural adjustment to CEE economies makes resistance futile and self-destructive, a conservative backlash against non-normative social developments provides an unfortunate, if widely-used, outlet for a frustrated citizenry eager to identify culprits for the evaporating existential security and astronomically growing social inequalities.

In addition to men who have forsaken their rightful dominant positions within the nuclear heterosexual family in favour of women – a trend discussed in Chapter 4 in relation to domestic abuse, Czech postcommunist gender norms and Shakespeare's *The Taming of the Shrew* – this conservative backlash also targets men who have chosen to withdraw from nuclear families altogether and to form alternative, exclusively male communities and partnerships. In their analysis of postsocialist gender formation in CEE, Susan Gal and Gail Kligman have pointed out that, in times of profound national shift, the definition of nationhood and citizenship hinges on culturally-sanctioned perceptions of sexuality. Sexuality here figures not only as an abstract category of identity formation and behavioural normativity, but primarily as a tool of policing ideologies of nationalistic normative reproductive practices that seek to dictate who procreates when, with whom and how often:

> 'citizens' are in many cases implicitly recognized as deserving of that title, and of the set of attendant 'rights' by their display of particular forms of state-sanctioned, legally acceptable, usually reproductive sexuality. And conversely, the reproduction of citizens is seen as beneficial, judicious, necessary for the future, while the reproduction of those not recognized as such – for instance, immigrants or stigmatized minorities – is seen, for that very reason, as dangerous, out of control, and polluting. (Gal and Kligman, 2000: 23)

In other words, the process of drawing ideological boundaries around a fledgling nation struggling with the aftermath of Soviet occupation draws on common perceptions of belonging and exclusion. Such perceptions, in turn, depend on an unhindered reproduction of its constituents, particularly at a time of precipitously falling birthrates. Non-reproducing homosexuals, for instance, can be charged with withdrawing their reproductive support for the nation, while homosexuals with a desire to reproduce within the legal framework of the postcommunist state threaten the newly redefined social policies on proper (essentially and traditionally conceived) heterosexual reproduction. Not surprisingly, in many Eastern European states, negotiations about citizenship inevitably resulted, as Gal and Kligman attest, in government promises 'to uphold the unchanging forces of natural gender order' (ibid.: 29).

Far from being concerned solely with the unorthodox gender roles homosexual men are perceived to create (considering, for instance, the stereotype of the softened, feminized, passive partner paired with an exclusively more masculine 'top'), the anxiety about homosexuality in particular and non-normative masculinities in general uncovers a deeper-seated concern with the future of a society no longer exclusively organized in neatly-defined heterosexual families and, more importantly, a concern with flawed citizenship at odds with a traditional understanding of core Western values. Since Shakespeare has served as a repository of theoretically universal Western human essence and homosexuality is perceived to be at odds with essential Western values, the resistance to positive representations of particularly male homosexuality in Shakespeare productions remains strong.

I. Homosexuality in the 'First Republic'

In explaining postcommunist attitudes towards homosexuality, particularly in the context of constructing essential Czech (masculine) identity as well as ideal citizenship, I take into consideration the historical development of historical cultural consciousness of this variant of non-normative masculinity in the Czech context.[2] Significantly, attempts to widen officially recognized iterations of masculinity within the larger cognitive framework of 'essential human nature' have historically coincided with political moments of national upheaval following recently gained independence from foreign domination when Czech legislators actively debated dimensions of legal national citizenship. The first documented movement for homosexual rights occurred in the 1920s and 1930s, shortly after the fall of the Austro-Hungarian Empire and the foundation of Czechoslovakia in 1918, even as the Czech lawmakers were finalizing the first independent Czech constitution in four centuries.

At the time, the movement included cultural elements recognizable in the postcommunist cultural context, similarly seeking to balance emerging Czech national identity with universal frameworks of Westernism and universalism. Both movements gathered force in the aftermath of a successful move for independence of the Czech lands from an imperial oppressor. In both cases, the Czech legislature was faced with the challenge to restructure the political and judicial systems in the colonial aftermath. In the 1920s, the Czech nation was in the early stages of recovery from four centuries of Austro-Hungarian colonialism (1620–1918), struggling to develop further and strengthen the concept of Czech national identity. In this politically exciting time, debates about the basic tenets of Czech citizenship abounded, examining the claims of various minorities to equal political rights and protection. Similarly, in the 1990s, Czechoslovakia and later the Czech Republic sought to shed the legacy of communism and return to a common understanding of national identity and citizenship. In these parallel contexts, the movements for homosexual rights challenged traditional assumptions about human natural sexual behaviour and demanded the recognition of homosexuality as a valid aspect of 'universal human nature', which would guarantee its inclusion in political and social legislation. In both cases, the activists for the cause as well as the legislators who responded to them were looking to appropriate and transcend established Western social and legislative models. Furthermore, both movements were hindered by a conservative resistance of the general population. Whereas the first movement was halted by the gradual onset of German National Socialism culminating in Nazi occupation of the Czech lands in the late 1930s, the postcommunist movement has as of yet no settled outcome.

We can gain a fairly good understanding of the agenda of the movement from the first Czech monograph on homosexuality, František Jelínek's *Homosexualita ve světě vědy* (*Homosexuality in the World of Science*), published in 1924.[3] It is worth reviewing at length, as it foreshadows not only the content of current debates about queer issues, but also the rhetoric of the Westernization of the Czech nation, morally acceptable science and the rhetoric of individual rights. Since the question of the origin of homosexual orientation seems to be at the centre of all discourse on the subject, as it can provide the basis for further medical and legal examination, Jelínek is careful to argue for homosexual inborn identity, not preference, claiming that 'it is impossible to acquire homosexual affections, as they are always inborn' (1924: 29, 46). Homosexuality itself is an '*intersexuální variac[e]*' (intersexual variation) within the spectrum of human sexuality, and hence it is natural (ibid.: 15).[4]

It is evident that Jelínek connects the issue of sexuality and sexual identity closely to the issue of nationality, and he sees tolerance of homosexuality as an integral part of an enlightened nation worthy of recognition by the established countries of Western Europe. Independent from the Habsburg rule for six years, the country, he argues, should strive to meet [Western] 'civilization standards', as well as the need for individual independence and rights. Jelínek himself sees his book as a tool of enlightenment for the citizens of the newly-established Republic, wishing that 'these pages travel to all Czech towns and hamlets, bringing everywhere the aura of truth to the benefit of the entire nation' (ibid.: 5). The Czechs as a nation, he claims, are already perfectly suited to understand and resolve the problem being 'naturally' prone to heed these 'civilization standard', since 'our people are in their constitution essentially thoughtful, downright revolutionary in their free thinking, accessible to liberal viewpoints and edification, in their centre is much good will, desire and appetite for light and truth, and that is why we are certain that even in our country the opinions on same sex love will thoroughly change' (ibid.: 60). The elimination of §129, which mandated heterosexuality, would allow Czechoslovakia to join the ranks of '*kulturních národů*' (i.e., cultured or civilized nations) such as France, Italy, Belgium, Romania, Russia, Japan, China, Portugal, Spain and the Netherlands (ibid.: 194). Accordingly, Jelínek draws heavily on his Western colleagues, primarily Hirschfield, Kraft-Ebbing, Kammerer, Lipschűtz, Freud, Moebius, Ellis and Carpenter.

Jelínek grounds his argument in the appeal to the individual as the integral unit of a cohesive civilized nation. Evoking individual human rights of homosexuals, he reminds his readers of their individual rights and the responsibilities of recognizing the plight of the 'Uranian' minority. He claims his conviction that every reasonable human being will understand the true plight of the sexual minority and join the battle for homosexual rights (ibid.: 194). His basic premise is that:

> an individual has the right of self-determination, and ... the state or the society do not have the prerogative to deny him one of the most sacred of human rights – the right to love according to one's nature – if he does not trespass on the rights of a third party ... It is high time to appreciate the fact that it is inappropriate to regard

> the foundation, which mysterious Nature lay in the cradle of a human minority, as criminality and vice, rather than a singularity for which no one can be deprived of a single even infinitesimal human right. (Ibid.: 3–4)

The 'high time' Jelínek evokes comes within the first Western European wave of discussions about homosexuality as a variant of personal identity and debates of the need for new legislation that would seriously take into account homosexual persons. Coming on the heels of the Czech emancipation from the Austro-Hungarian Empire in 1918, Jelínek's appeal to this 'universal' or 'civilized' trend seeks to alter traditional Czech understanding of citizenship to include the homosexual minority on an equal footing.

The central arguments of Jelínek's appeal – the insistence on homosexuality as a natural, inborn trait; the cultural cleansing of the concept of homosexuality of socially preconceived notions such as pornography, paedophilia and perversity; and the parallel between a free Western national identity and freedom of the homosexual individual – resurface repeatedly throughout Czech history, particularly in moments of increased cultural freedom. The appeal to Western models, both scientific and cultural, signifies in particular its attempt to define Czech cultural identity as part of the Western tradition. And yet, while for one camp the protection of homosexual rights seems to evidence advanced mode of civilization equally respectful of all its citizens, the majority seems more focused on the coercive qualities of the recommendations of the established West, so that the 'equality' clauses within the EU recommendations are interpreted as attempts to sully – and therefore weaken – the efficacy of the otherwise ideologically unified perception of essential masculine citizenship.

II. Communism and homosexuality

Before it claimed control of the Czech political sphere in 1948, the Communist Party was one of the most staunch supporters of homosexual rights. Claiming that in a communist state all minorities would be equal, the Communists criticized the capitalist bourgeois system for demonizing homosexual men and women. Consequently, the Party proposed the dismantling of the controversial §129, earlier criticized by Jelínek and other activists in the movement for homosexual rights. Following its sweep of political power, the Party proved reluctant to rearticulate normative boundaries of citizen subjectivity. The punitive law remained part of the Czech legal code until 1961, though the Communists renumbered it to §241 in 1950. In 1961, new legislation specified that homosexual acts were basically legal, provided that they were consensual, between persons older than 18 and with no payment involved. Apparently, however, this formulation was still easily manipulated, since even an invitation to dinner could be seen as a form of 'payment' and a sexual relationship between two 17-year-old boys could have been interpreted as illegal as soon as one of them turned 18.[5] Though homosexual acts were technically permitted, the public display of same-sex affection, discussion of homosexuality or its representation in art was not.

Within this Orwellian universe, despite the public rhetoric of equality, traditionally-defined normative subjectivity remained the centrepiece of socialist selfhood. Though homosexual behaviour was thus legalized and homosexuality itself became more visible, cultural and even legal acceptance was not achieved. Historian Jiří Fanel argues that in communist Czechoslovakia 'ruled the idea of the official establishment that homosexuality [was] "not compatible" (which was the magical formulation) with the morality of the socialistic man' (2002: 434). As Pavlína Janošová attests, the state prevented homosexuals from gathering and it silenced attempts to inform the wider public. Public display of affection could be prosecuted as *'jednání vzbuzující veřejné pohoršení'*, that is, 'behaviour evoking public scandal'. Publication of same-sex classified ads was therefore prohibited until the end of the 1980s. The consequences, Janošová argues, were 'a generally low level of information available to public attended by a relatively high occurrence of prejudice and negative views of our society. Some members of the minority understandably refused to live according to their orientation and married according to their reason' (Janošová, 2002: 48).

III. Homosexuality in the postcommunist Czech Republic

The developments in the postcommunist culture were greatly influenced by the society's rapid and forceful shedding of the totalitarian, compulsorily communist *modus operandi* in favour of returning to the focus on the individual as the pivotal point of society. Contemporary rhetoric of individual power and of Czechs as a 'civilized' and already 'Western' nation is reminiscent of the arguments brought forth by Jelínek in 1924, this time around fuelled further by the urgency to jump on the bandwagon of Western Europe on both the economic and political levels to meet the standards of the European Commission to join the EU. Seventy years after the publication of Jelínek's book, on 2 February 1994, the European Parliament of the EU advised its member states to alter their legislation so that homosexual partnership would fall into a protected category, and voted by a close majority for the possibility of adoption for same-sex couples. Many of the member states had by then already legislated anti-discrimination laws.[6] Though not a requirement for admission to the EU per se, such 'advice', together with similar guidelines about racism as well as sexism clearly registered with Czech society, which resulted in rapid, if mostly surface changes in Czech legislation.

Seven months after the Velvet Revolution, in June 1990, various small organizations came together under the collective umbrella of SOHO – Sdružení Občanůs Homosexuální Orientací (the collective of citizens with homosexual orientation) – and elected its first president, Jiří Hromada. SOHO became immediately politically and socially active, quickly accomplishing a number of legislative changes. Over the decade, there were three attempts to legalize registered partnerships: in 1995, 1998 and 1999. On each occasion, the legislation failed by a narrow margin of two or three votes. On 1 January 2001, the workplace anti-discrimination law, sponsored by the Social Democrats, was passed. In the last national elections in June 2002, Czechs elected by an overwhelming majority the ČSSD – Česká Strana Sociálně Demokratická (the Czech Social Democratic Party) – that initially included in its party election programme the legalization of registered partnerships by the end of its term in 2006. In keeping with the ambivalent stance of the culture towards homosexuality in general, the clause was removed immediately after the elections at the bidding of the KDU-ČSL (a coalition of the Christian-Democratic Union and the Czech Party of the People), a more conservative political party with which the ČSSD created a government coalition to secure a majority in the senate.[7]

Information about homosexuality, nearly non-existent under communism, became more readily available after 1989. Still, the number of publications on the issue is small, and most of the offerings are issued by publishers in the field of medicine. Pavlína Janošová, in her study of homosexuality in the Czech Republic, allows that the population is far more informed now than it had been prior to 1989, though she points out that the situation is comparable with the position of gays and lesbians in Western societies 20 years ago (2002: 91).[8] This reluctance to embrace homosexuality is reflected in a larger commitment towards ideological preservation of dominant models of masculinity, seen as a far more productive and effective foundation of the Czech postcommunist society.

IV. Performing non-normative masculinity in postcommunist Czech art

The partial acceptance of non-normative masculinity seems to be supported by portrayals of homosexuality in the postcommunist popular culture. Relatively quickly after the fall of communism, for instance, Hudební divadlo Karlín (the Musical Theatre Karlín), which is well known for performing famous US musicals such as *Some Like it Hot* or *Hello Dolly*, successfully produced two musicals with explicit homoerotic and homosexual themes. A Czech version of Jerry Herman and Harvey Fierstein's *The Birdcage* opened on 30 May 1998 and in the 2001–2 season the theatre staged Henry Mancini and Blake Edwards's *Victor-Victoria*.

The reviews of *The Birdcage* reveal the novelty of the issue, previously absent from active Czech cultural consciousness, as well as a level of willingness to consider the issue seriously. Branding *The Birdcage* 'šokující muzikál' (a shocking musical), the reviews revolved almost exclusively around the issue of homosexuality. Calling the production a 'test of the audience's tolerance',[9] the reviews repeatedly pointed out the connection between art and politics,[10] commenting on the show's and the cast's attempts to promote tolerance for queer issues and persons. With the minor exceptions of those who need to point out that the actors playing the roles of homosexual characters were themselves heterosexual,[11] the press took an accepting standpoint on the issue, generally challenging the audience to engage with the subject and 'think about not judging something that is different from what is our own, and let others live'.[12] As part of the review process, many of the critics promote registered partnership legislation debated at the time in the parliament.[13]

In contrast, the commentaries on *Victor-Victoria*, the second production 'of the kind', are far less focused on homosexuality; instead, they address non-normative masculinity via pronounced focus in the title cross-dressing role of Victor/Victoria, thoroughly debating the cultural ramifications of a woman's success in performing masculinity. Accordingly, the majority of the reviewers interviewed Michaela Dolinová, the actor playing the title role, about the presumed intricacies of playing a male character. Tellingly, while modestly accepting seemingly stunned praise of her performance, Dolinová frequently underplays the implications of her performance of the masculine Victor by reiterating her unwavering commitment to heterosexuality, marriage and motherhood. Political and social commentary on the part of reviewers is absent. Instead, the reviews emphasize the musical's established success as a Hollywood blockbuster, a spectacular show worthy of the modern Czech audience's attention.

As successful as these musicals were with numerous audiences, they were firmly shackled to the category of trash popular culture. This label inevitably undermined the gravity with which the productions explicitly sought to enter the field of political debate and influence the cultural conscious of their spectators. The theatre professionals responsible for the production of both *The Birdcage* and *Victor-Victoria* sought to minimize this connection and repeatedly addressed the issue in press releases, interviews and programme notes. In an interview printed in the programme for *Victor-Victoria*, for instance, director Robert Dubský reacted to the widespread criticism of the current increase of the number of musicals performed in Prague at the reported expense of traditional 'serious' theatre productions. In his headnote to the production, he compares current theatre with the theatre of the 1970s. In the 1970s, he argues:

> people had other worries, but fewer existential ones than today: they were more willing to engage with problem plays. If current viewers do not attend such productions in numbers comparable to audiences at 'insipid' playettes, where is the error? In the viewers? the plays? the authors? the actors? … Viewers' interest is created freely and spontaneously, it is impossible to dictate what should or should not be performed, what should or should not be liked. If we are to offer something to the viewer as creators, if we are to somehow reach her/him (whether a play or a musical has

some 'message' or 'just' catchy melodies), it again depends only on the viewer what he/she will choose and what, therefore, has a chance to survive. It is a question of the market and of theater as a part of this market.

Dubský's claims are relevant not only in their explicit articulation of theatrical and cultural developments over the span of four decades, influenced by the political and sociocultural changes, but also in the new consumerist context of theatre as business. Under socialism, the question of 'theater as a part of [a] market' was unthinkable or, more accurately, it was not to be openly articulated. The official stance maintained the power of the state to censor and to dictate appropriate public tastes and subsequent cultural consumption. It is precisely this realization – that satisfied paying audiences are crucial to a venue's survival – that has uncomfortably entered the equation in the postcommunist era. Stripped of their state subsidies and forced to depend on in-house revenue and external sponsorship (often directly proportionate to a theatre's success and popularity), theatre directors and producers determined repertoire with an eye to its likely financial success. 'High art', usually perceived as profoundly enlightening but assumedly indigestible in its inaccessibility has, thus battled with popular theatre for revenue and spotlight. In this context, theatres with traditional 'high art' repertory needed to find new ways to sell culture. In contrast, theatres with traditionally lighter fares sought to gain cultural prestige by attempting to undermine the traditional cultural barriers that have separated the 'low' from the 'high' and claim at least nominal cultural equality.

The producers of *The Birdcage* and *Victor-Victoria* contextualize their appeal to the power of the discerning Czech consumer, free to choose entertainment and determine the material value of available art, in the immense success of the musicals on Western stages as well as in Hollywood. As such, what is presented on stage is not as much a musical, or an example of a popular artform new to the Czech context, but rather a sensationalistic sampling of established popular spectacle that the West has found worthy of cultural consumption. This perception of the show as substitute-for-the-original is further buttressed in each production's close adherence to the Hollywood performance template – in the instance of *Victor-Victoria*, for example – in duplicating the content, costumes, form and setting presented by the 1982 film version starring Julie Andrews. Since Dolinová's masculine portrayal of Victor closely follows the visual model of Andrews in body movement and mannerisms, any challenge to the notion of natural Czech masculinity and its performance is relegated to the US template rather than to the Czech copycat production. The connections to Western spectacular success were successfully marketed in the production of *The Birdcage* as well: the programme notes for *The Birdcage* recount the musical's success on Western stages – particularly Broadway – and in film, justifying both the genre and the content. Some reviews underline the 'grand [Western] style' of the productions, and others explicitly compare it with other 'American' shows.[14]

Thus, in their passionate act of redeeming their musicals from the taint of 'popular culture', the producers succeeded in relegating these two pieces into the realm of the thoroughly 'foreign' that connects only ambivalently with the cultural context of the Czech audience. Not even a direct appeal to Shakespeare, in lieu of an alternative connection to 'high(er)' art, erases this distance. Evoking Shakespeare as a direct predecessor of the subject matter of the musicals, for instance, the *Victor-Victoria* programme argues for the timelessness of the central cross-dressing role as a continuation of Shakespeare's tradition:

> The musical *Victor-Victoria* belongs among works which include one of the most interesting moments in dramatic art: a cross-dressing role – here a beautiful young woman, who pretends to be someone else, who can dance, sing, act and evoke love – her position in the narrative becomes more entertaining, witty, and expressive. The disguise of the main characters often drives the entire plot. Doubling is an ancient theater trick: let us remember Shakespeare's comedies such as *Twelfth Night, As You Like It,* [or] *The Merchant of Venice.*

Calling on the cultural capital of Shakespeare, the producers thus seek to justify both the format of delivery for their spectacles and its content that potentially challenges postcommunist structures of Czech identity, be that challenges to the 'natural' dimensions of normative masculinity in the figure of Victor/Victoria or explicit homoeroticism in both musicals. In insinuating that non-normative masculinities are already present in Shakespeare, the producers claim to build explicitly on a Shakespearean foundation to address similar issues reconfigured for the cultural framework of postcommunist society; as Ladislav Županič, the director of Karlín, proclaims: 'We are bringing *The Birdcage* at the exactly right moment as a contribution to the discussion about registered partnership. Legalizing of partner relationships is extremely important for those people. And the main function of theatre, after all, is to mirror life, as Shakespeare used to say.'[15]

V. Postcommunist (Shakespearean) masculinity

That the link of Shakespeare to the popular genre of the musical – and its culturally daring subject matter – fails in the Czech context becomes evident in a survey of representations of non-normative masculinities in postcommunist Shakespeare productions. A comparison of a recent performance of *The Merchant of Venice*, a genealogy of Czech productions of *The Sonnets* and two representative productions of *Twelfth Night* demonstrates that the relatively positive reactions to representations of non-normative masculinities in the popular media speaks only of a partial acceptance and tolerance within the postcommunist Czech culture. The fact that the audiences far less readily accept the positive portrayal of homosexuality at the site of the tradition – the quintessential Western civilized space of Shakespeare – suggests a tension between different arenas of cultural productions that undergird competing cultural conceptualizations of postcommunist Czech identity and citizenship.

The portrayal of masculinity in Shakespeare evokes two competing perceptions of Westernism. One is the mythical site of tradition, which supposedly contains repeated iterations of transcendental essence of all humanity. This essence is also reportedly – and conveniently – embodied in Shakespeare, though it is important to distinguish here between the essential, or 'true', Shakespeare and the Shakespeare who has been bowdlerized by inappropriate handling by scholars and theatre professionals. It is this tradition, carefully policed by the cultural apparatus (addressed in Chapters 2 and 3), that has provided the ambiguous-but-crucial 'truth' evoked as the cornerstone of postcommunist Czech nationhood. This transcendental template of humanity is frequently perceived to be at odds with the other version of Westernism, which has been crossing the national boundaries from the established West in the form of Western media, entertainment and legislation enforced by the EU. Profoundly culturally committed to the one, but generally wary of the other, Czech culture struggles to find its true roots and modern cultural identification. The tensions surrounding postcommunist variants of masculinity, particularly homosexuality, exemplify this ideological rift: while the increasing tolerance for the portrayal of homosexuality in popular culture suggests some willingness to adhere to EU-suggested models of citizenship, the spotty implementation of new legislation and the ongoing invisibility of homosexuals in society suggest that the limited tolerance for homosexuality is a tolerance of a surface and fleeting trend by no means essential to transcendental identity. Denying homosexuality its place in the traditional space of the 'transcendental genius' of Shakespeare provides the resistant portion of the Czech population with viable hope that 'this' – like the 1970s fashionable trend of bell bottoms – will 'pass'.

One of the first Shakespearean challenges to the boundaries of normative masculinities came from the otherwise rarely performed *The Merchant of Venice*,[16] directed by Michal Dočekal at Komedie, a small repertory theatre well known for radically alternative productions of canonical texts. First performed in 1997, the production lasted in the theatre for 120 performances over a period of four years. Shying away from foregrounding homosexuality as the only variant of non-normative masculinity in question, the production played with a variety of masculine models that vied for symbolic power and efficacy throughout the staged narrative. Transposed into a modern context, the production addressed the difficult aspects of postcommunist identity construction in a cultural context increasingly stripped of recognizable ethical and moral markers. This tension surfaced on multiple axes of the play; it marked the generational conflict between, on the one hand, Antonio and Shylock and, on the other hand, the younger Venetians, Jessica and Portia's household. It equally haunts the referenced cultural contexts of exploitive trade in and abuse of narcotics, unbridled consumerism, chaotic immigration and surfacing racism. In individual interviews, Dočekal, as well as Miroslav Mejzlík (Antonio), for whom the production was a form of a public coming-out, spoke of the decision to 'slip in' homosexuality rather than pivoting the production on it to make a political statement. For some in the cast, like David Matásek (Graziano), Antonio's homosexuality was too subtle to notice within generally 'sexually unclear' relations between the Venetians as a group.

In Dočekal's production, the quasi-aristocratic Venetians of Shakespeare are portrayed as a highly-functioning, criminally-active, well-dressed mafia of young slick men surrounding an older, more experienced Antonio, whose 'merchant' status seems to come mainly from a profitable drug trade and violent crime. His attendants – Salenio, Solario, Graziano and Bassanio – visibly compete for his affection, money and favours, which seem to create the centrepiece of their existence devoid of any other aspirations or moral referents. The explicitly homoerotic relationship between Antonio and Bassanio, though seemingly sincere on Antonio's side, can be read as pure exploitation by Bassanio, who is primarily interested in an acquisition of material resources. Their relationship is further complicated by the appearance of the wealthy, experienced, confident and not-easily ruffled transsexual Portia (see Figure 5.1), who presents a

Figure 5.1 David Prachař as Portia. *Kupec Benátský* (*The Merchant of Venice*), dir. Michal Dočekal, Divadlo Komedie.

new source of material wealth and prestige to Bassanio once Antonio's stock plummets. Portia and Bassanio's subsequent marriage is clearly based on a socioeconomic interpersonal agreement rather than mutual affection, and no amount of ardent Shakespearean wooing language can disguise the characters' physical indifference towards each other; their polite coldness contrasts sharply with the behaviour of the other couple – the eloped Jessica and Lorenzo – who frequently display the physical dimensions of their sexual desire on stage.

The multiplicity of non-normative identities on stage thus seeks to address the context of postcommunist society unhinged by the application of neoliberal frameworks that promote personal enjoyment over civic responsibility at the expense of overall societal organization. The resulting chaos, wherein the relationships of a mafia replace familial ties of the central characters, speaks to a range of contemporary issues, from the dangers attendant on neoliberal deregulated market structures benefitting the few at the expense of the society to the attendant criminalization of the population unable to tap into established social frameworks of existence, the corruption of the justice system, the discord between an older, religious and conservative generation (represented here by Shylock) and the younger postcommunist generation focused solely on profit and entertainment (the Venetians) and, finally, the consequences of the teeming centrality of consumerism to self-identification, visible in the transformation of the impressionable and naïve Jessica into a superficial peroxide blonde with no other existential goal than excessive consumption (of material goods and of sex). The result is a rather dire critique of the postcommunist society as being in danger of utter economic and social collapse, where sexual licence (which ambiguously includes homoerotic behaviour) signifies the growing self-absorbed individualism championed by neoliberal consumerist values. Within this context, Antonio's desperate love for Bassanio becomes lost in the fray, unclear as to its ultimate significance within the larger commentary on deteriorating masculine models of postcommunist identity.

Despite – or perhaps because of – this multi-levelled currency, reviews of Dočekal's *Merchant* tended towards the negative. Though individual critiques focused on various subplots, the reviewers overall agreed that Shakespeare's comedy was stripped of much of its complexity and humour in favour of politicizing the content. Speaking to the dilemma of contemporary tendencies towards political correctness that would force theatres to invent reasons for Shylock's hatred of Antonio,[17] all reviewers faulted the production for its contemporary setting that reportedly failed to do justice to Shakespeare's renaissance original and, especially, for exaggerating the immorality of the Venetians. For some, like Alena Urbanová, the substitution of questionable homosexuality for the traditionally interpreted bounds of friendship between Antonio and Bassanio was unforgivable.[18] In addition to pointing to parallels between the production and the alarming consumerist trends in postcommunist Czech society, most reviewers focused on the non-normative spectacle of Portia. This, arguably the first transsexual performance in mainstream theatre, was widely read by the reviews as further evidence of the Venetians' moral depravity, already established by the excessive sexuality which includes the homosexual relationship between Antonio and Bassanio. Most reviewers found Portia's non-normative identity disturbing and inappropriate, further disrupting the traditional qualities of Shakespeare's comedy. Unlike contemporary productions of US musicals, where non-normative masculinity is greeted by excited curiosity that simulates public tolerance and understanding, its equivalent – however critically performed – in the heart of the established cultural canon, Shakespeare, earnt unequivocal derision.

In contrast to the sensation of transsexual Portia, *The Sonnets* have traditionally been offered as an intellectual and romantic dainty for the 'cultured' audience. Since 1945, the press has reviewed six public productions of *The Sonnets*. Most of these productions have taken the form of stylized and 'cultured' poetry readings, with one or two actors reading selections of the poems accompanied by a Renaissance musical instrument or a small ensemble, usually emphasizing enjoyment of poetry and a collective celebration of Shakespeare's transcendental genius rather than a critical interpretative response to the work. For instance, in August 1975, Lyra Pragensis, a prestigious literary club, sponsored a sold-out *Sonnet* reading by Milan Friedl, accompanied by a performance of Dowland compositions. Four years later, a similar club, Reduta, offered a reading by two prominent actors, Jindra Šolcová and Jan Přeučil, to the accompaniment of Collegium Flauto Dolce, a period ensemble. The next performance in 1986, again organized by Lyra Pragensis, was similarly complacent, so much so that reviews commented that the performance failed to 'address the ancient quarrel about the [sexual] interpretation' of the texts.[19] While perhaps textually inaccurate in their erasure of gendered pronouns, which heterosexualized the sexual desires expressed in Shakespeare's poetry, these performances buttressed the transcendental heteronormative themes allegedly central to the work of Shakespeare to whose timeless genius the Czech cultural elite here collectively expressed due appreciation.

Postcommunist productions of *The Sonnets* have taken a somewhat less celebratory and unquestioning approach to Shakespeare to provide more reflective material for the audience. While a 1990 Lyra Pragensis production seemed to continue in the traditional reading vein, it nonetheless provoked a reviewer to comment that the performance made obvious the significant connection of Shakespeare's poetry with the 'cultural development of our country'.[20] Another reading in 1992, while continuing the venerable tone of previous *Sonnet* performances, addressed Shakespeare's genius through an extended discussion of the difficulties of capturing the poems' beauty and complexity in Czech translation. Organized again by Lyra, the evening's host was a significant Shakespeare scholar and Professor of English, Zdeněk Stříbrný, formerly silenced by the communist regime, who interviewed prominent Czech translators of *The Sonnets* about their approaches to and interpretations of Shakespeare's poetry. This interchange was interlaced with readings of the translations by a whole spectrum of well-known Czech actors.

The most innovative performance of *The Sonnets* prior to the fateful Montmartre production came in 1996, when 'The Labyrinth', a small theatre lab, crossed the boundary of a poetry reading to engage four actors and four singers to perform a selection from the collection. It also entered the unfamiliar waters of addressing both the seemingly difficult sexual relationship between the speaker and the 'dark lady' as well as the initial admiration and worship of the youth. The press reactions were mixed. Few reviews were favourable, commenting that 'love is love, regardless of who loves whom, whether it be men loving men, women loving women, or the other sex'. Others skirted around the controversial issue of homosexuality and focused on the heterosexual aspects, claiming that '[a]ll that is beautiful should procreate',[21] or summarized the performance as one that emphasized the importance of the intellect over the emotions, so that 'the verses addressing the young man appeal primarily to his duty to marry', whereas the poems addressed to the woman are seen as appropriately playful. Yet other reviewers implicitly criticized the production's attention to multiple objects of affection, citing a lack of overall coherence of narrative, artistic expression and/or direction.[22] The director Karel Kříž himself carefully avoided addressing the issue, shielding the production by describing the persona of transcendental Shakespeare as a universal 'genius' who presents us with the challenge of 'breaking the code' he sends across centuries.

The Montmartre production I described at the outset of this volume contrasts the most with the traditional interpretation of *The Sonnets* as a poetry reading. The two male actors, Kreuzmann and Trtík, emerge from the rubble of their scene minimally clothed to act out a narrative of love, desire, refusal and, at the end, acceptance. This narrative suggests an intense search for intimacy, for identity and for truth – especially in the despairing oscillation between desire (of a man for another man) and the socially proscribed path (procreation and children). The homeless setting of a landfill intensifies the action: the dialogue at times becomes violent, both verbally and physically. The end offers a peaceful resolution, where both realize that love between them is possible, and that it lives on in the poetry itself: they recite the last sonnet (Sonnet 76)[23] in echoes – together yet on their own, as if they were coming to a joint and yet independent realization about the nature and inherent value of their mutual affection.

In a group interview, the actors and the director reported that, especially when they began performing *The Sonnets* in 1999, the middle-aged audience used to traditional Lyra Pragensis productions was appalled, thinking that the production disproportionately underscored and even propagated homosexuality. The actors and the director pointed to Shakespeare's text as the source of the homosexual relationship and claimed that their goal was to speak about love in general – love that oversteps boundaries of gender, one that binds together people regardless of who they are. They also sought to connect *The Sonnets* to the damaging neoliberal practices implemented in Czech postcommunist society. Pavlík claimed:

> I feel hope here; we tried to convey the direction in which the globalized world is going, this world of complete lack of interest for another human being. A person realizes the depth of feeling for another, and simultaneously realizes that he [*sic*] is alone in

the world, and that the only path is to live one's life honestly and fill oneself with true genuine love which then becomes eternal.

Whereas the numbers at Hudební divadlo Karlín suggest that Czechs of all ages flock to see *Victor-Victoria* as a pleasingly exotic and fashionable production of non-normative subjectivity, the experience of the artists at Montmartre painfully points out a dominant refusal to accept homoeroticism in the perceived core of the Czech cultural canon.

VI. *Twelfth Night*: a study in postcommunist gendered identity

Recent performances of Shakespeare's *Twelfth Night* are neither as scandalously spectacular as Czech versions of US musicals nor as unambiguously challenging as the 2001 production of *The Sonnets*. As such, they exemplify recent concerns with new definitions of citizenship refracted by a Shakespearean lens. The two recent productions I will discuss – Enikő Ezsenyi's *Twelfth Night* in the National Theatre (2001) and Viktor Polesný's *Twelfth Night* at the 2005 Summer Shakespeare Festival – exemplify this conflict and, further, reflect an intriguing divide along the parameters of genre and venue in their representations of masculinity and, in particular, the relationship between Antonio and Sebastian.[24] Prominent, reputable and well-attended theatres, I found, though eager to present a 'contemporary' Shakespeare to affirm that they continue to move 'with the times', seem to shy away from exploring the intensity of the characters' emotional commitment to each other, downplaying their interactions and editing their lines. Small theatres, perhaps eager to draw audiences with new material, seem far more likely to take risks in presenting unorthodox readings of Shakespeare's plays, but even there much care is taken that such productions relegate non-traditional masculinities into the realm of the exotic, the fantastic and even the ridiculous.

The source text of *Twelfth Night,* preoccupied with multiple plots of coupling of eligible bachelors and bachelorettes, lends itself particularly well to the explorations of various forms of masculinity. Its multiple, sometimes overwhelming references to and examination of manhood suggest that the worth of individual male suitors – and their suitability for successful partnership – rests on the kind of masculinity they inhabit. Duke Orsino, though seemingly the most eligible of all bachelors – as his beloved lady Olivia admits herself (I.v.227–31)[25] – too closely approximates the stereotype of a soft, effeminate aristocratic courtier, more interested in a courtly embodiment of the suffering lover than in his professed beloved. Sir Toby's lack of resources, coupled with his love of drink, nearly disqualifies him from the eligible pool. Wealthy Sir Augecheek is weighed down by slow wits. Malvolio is hindered by his servant status and undue ambition. Finally, Cesario suffers from his professed eunuch identity that disguises the cross-dressed Viola. All of the nominally male characters fail in the ultimate test of masculinity in their shortcomings as *soldiers*. The play categorizes their swordsmanship – the ultimate marker of the degree of the phallus they possess – as lacking, which results in various spectacles of public shaming.

The two bachelors who do satisfy – at least initially – Renaissance requirements of eligibility in manhood are Antonio and Sebastian, who never hesitate to protect their physical persons, symbolic honour and each other by valiant sword-wielding. These two characters also like each other to a degree that reminds us of a relationship between lovers. In particular, Antonio reveals that for three months 'no interim, not a minute's vacancy,/Both day and night did we keep company' (V.i.83–5). Deeming himself bewitched (V.i.64), Antonio does not hesitate to risk his life following Sebastian to Orsino's hostile territory, since Sebastian's absence would 'murther' him (II.i.35). As he sums it up, he gives Sebastian his 'love, without retention or restraint, all his in dedication' (V.i.69–70). Sebastian corroborates this account, proclaiming that, even after he has married Olivia, any time apart from Antonio 'rack[s] and torture[s]' him (V.i.203).

Significantly, Antonio's exclamations of same-sex love parallel, and therefore in the world of the play equal, Cesario's heteronormative remarks uttered as he/she marches to seemingly certain death at her beloved Orsino's command (V.i.115). Similarly careless of preserving his/her life, Cesario proclaims that he 'most … willingly … a thousand deaths would die' for him whom he loves 'More than I love these eyes, more than my life,/More by all mores than e'er I shall love wife' (V.i.122–6). Aptly, this speech, etched into Orsino's memory, ultimately brings Cesario the husband he/she desires once he/she sheds his/her gender ambiguity and turns woman again (252–3). Unlike Viola, Antonio fails to earn a husband by his proclamations of love for Sebastian, though he perhaps retains the position of Sebastian's bosom friend. After all, the ending of the play does not provide a clear delineation of any of the characters' emotional trajectories, and Antonio's textual silence could equally signal either his dismissal from the play (a common interpretation in current productions) or his complacent figuration as Sebastian's primary male companion not in competition with Sebastian's marital duties to Olivia.[26]

The potential for the longevity of Antonio and Sebastian's relationship past the closing of the play is further strengthened by Renaissance discourses of friendship and masculinity, wherein same-sex affection between two 'bosom' friends, often expressed in language that we may consider erotic, functioned as a pinnacle model of true intimacy. In a world of innate inequalities, where all are bound to frameworks of vertical service, a voluntary, mutual – and sometimes sexual – relationship between men of similar social standing marks one of the few venues for a genuine connection. Contrary to twentieth-century terminology, the label 'effeminate' in the Renaissance applied to a man who spent – or desired to spend – a disproportionate amount of time away from the masculine pursuits of arms and exercise in the company of women (like Orsino). In contrast, a man who spends his time in male company has an opportunity to grow to true masculinity and arms. As Stephen Orgel pointed out in *Impersonations*, Antonio is one of the exemplary 'real' men in the Shakespearean canon, a position buttressed by his complete independence from women (1996: 81). Sebastian's continued allegiance to this 'fighter/pirate – and lover of boys' may balance Sebastian's initial subservience to Olivia's wishes and ensure the preservation and further development of his masculinity.

Examining contemporary popular knowledge of these characters, Cynthia Lewis has argued in *Particular Saints* that Renaissance audiences most likely greeted Shakespeare's Antonio and Sebastian with comfortable recognition of a stereotypical pair, Antonio and Sebastian, who functioned as a model of steadfast friendship. Drawing on the medieval and Renaissance hagiography of St Anthony and St Sebastian, both of whom eventually became patron saints of a complex network of hospitals caring for those suffering from painful incurable diseases ranging from St Anthony's Fire to syphilis, Lewis argues that:

> By the late sixteenth century, the names Antonio and Sebastian alone must have provided a literary-theatrical shorthand to an audience that, upon hearing them, would recall a host of possibilities for and expectations of these characters. Most significantly, the audience would have instinctively associated Antonios with extravagant love and with the difficulty of expressing such love as human beings in the human sphere, which is bound by constraints. (1997: 15–16)

In other words, a theatrical pairing of Antonio and Sebastian would have immediately signalled to the audience an imminent representation of characters who struggle to express their literally transcendental love for each other within acceptable earthly boundaries, limited both by human nature (bodies) and nurture (acceptable cultural norms).

Modern directors of *Twelfth Night* productions face a paradoxical dilemma of interpreting the play in a context in which perceptions of masculinity, individuality and sexuality have changed so profoundly that a one-to-one correspondence with the Renaissance meaning of Shakespeare's text is no longer possible. Because of the current political climate wherein both masculinity and homosexuality are contentious at best, any interpretation of Antonio and Sebastian's relationship in a contemporary production becomes inevitably political. Individual performances thus reflect an uneasy intersection of a director's reading of the text, his/her willingness to challenge the text's history on the stage and in literary criticism, and the degree to which he/she pays homage to the contextually-defined cultural capital of Shakespeare as the genius of 'all time'. While no interpretation can approximate the complex Renaissance understanding of masculinity in *Twelfth Night,* these performances speak volumes about current tensions between traditional and redefined views on what makes or breaks a 'true' man. Further, in the Czech Republic, the questions attendant on new definitions of masculinity make visible the minute details of the process of constructing a new 'Western' identity, an identity (perhaps wishfully) taken for granted in most of the 'developed' world.

Enikő Ezsenyi's production at Stavovské Divadlo, one of the stages of the Czech National Theatre, though in many ways presenting an alternative view of the play, let the homoerotic themes present in the text lie by the wayside. Contemporary in setting, which happens to be a run-down wintry country farm, this version comes across as a spectacle of stunning stage design and costumes. It underscores the themes of wealth, leisure, relaxation and, most notably, sex. Almost every instance of Shakespeare's sexual innuendo seems to have been not only duly noted, but also transformed into a corresponding (mostly non-verbal) stage action. Figure 5.2, for instance, captures the production's unambiguous take on the nature of Sebastian and Olivia's relationship ensuing from Olivia's confusion of Sebastian for his identical, cross-dressed and much-beloved twin Cesario/Viola. Where the source play may leave some ambivalence about the hasty marriage that ensues as soon as Olivia can physically capture her perceived object of desire, Ezsenyi's explicit rendition of their

Figure 5.2 Martina Michal Slaný as Sebastian and Válková as Olivia. *Večer Tříkrálový* (*Twelfth Night*), dir. Enikő Ezsenyi, Stavovské Divadlo.

Figure 5.3 Vilém Udatný as Antonio and Hana Ševčíková as Cesario. *Večer Tříkrálový* (*Twelfth Night*), dir. Enikő Ezsenyi, Stavovské Divadlo.

marital consummation leaves no doubt about the rightness of their heterosexual coupling. The only sexual innuendos that are conveniently omitted are those suggested between the potential same-sex couples, Olivia and Cesario and Antonio and Sebastian. Again, elaborate non-verbal action guides the audience in viewing the possible same-sex connection as volatile and violent (see Figures 5.3 and 5.4).

In this production, Antonio appears more interesting as a wealthy, hypermasculine but ultimately socially deviant, gun-toting blade-swinging drug lord than as an object of or subject to affection towards Sebastian who, in turn, is freed from the emotional complexity of multiple commitments. Antonio functions as a mere detached spectator to Sebastian's commitment to Olivia who subsequently blends into the celebratory wedding fray of Act V. The relationship between Antonio and Sebastian does not serve any particular purpose in

Figure 5.4 Michal Slaný as Sebastian and Vilém Udatný as Antonio. *Večer Tříkrálový* (*Twelfth Night*), dir. Enikő Ezsenyi, Stavovské Divadlo.

the play beyond moving the plot forward, thus effectively erasing homosexuality as a positive variant of normal male identity, behaviour and sexuality.

The context of Ezsenyi's *Twelfth Night* makes this erasure profoundly political. Though this production seems carelessly frivolous and has been dubbed repeatedly as such by disenchanted reviewers,[27] it cannot help taking a political stance by its mere presence at the National Theatre, where the play has a significant stage history. Shakespeare's *Twelfth Night* prematurely closed a season on 31 August 1944, the night before the fascist regime shut down Czech theatres to repress subversive nationalistic expression. This closing performance is still remembered as teary and emotional (Přibyl, 2001b: 80–1), anchoring the endangered Czech nation in the higher transcendental values of Shakespeare's text. Once the War ended in May 1945, *Twelfth Night* rapidly returned to the stage on 20 June 1945, barely a month after Soviet tanks rolled into Prague. In the subsequent communist years, the play served further political purposes, centring around Malvolio's failure of manhood as he surrendered himself to the service to established institutional and political powers and thus failed to grasp the true essence of humanity (ibid.: 84). This view of Malvolio quickly lost its salience in the postcommunist era, as the old political framework for understanding the play crumbled together with the old political regime, and society entered the general confusion of the 1990s. As the first *Twelfth Night* to be produced at the National Theatre in 36 years and the first since the fall of communism, Ezsenyi's production negotiates the uncertain ground of a nominally 'post-political' as well as expectations of a worthy interpretation of a politically significant Shakespeare play.

The confused and piecemeal production, we could argue, ultimately does succeed in portraying the fragmentation of Czech culture. Unsettlingly disjointed, the narrative allows for little character development, instead offering a series of loosely organized, independent vignettes, crafted to amaze by their theatricality, slapstick acrobatics and unconventional exposure of body parts. In this context, Sebastian's simplified model of masculinity, which denies his character emotional complexity, indicates only one of the shortcuts the production makes in trying to grapple with the instability of the Czech postcommunist culture. The presence of such a fractured production of Shakespeare, a repository of essence of universal human nature, in the Czech National Theatre, a space traditionally reserved for artistic explorations and validations of the Czech national spirit, implies a current lack of coherence of the essence of Czech nationhood in general and its building blocks – individual identities – in particular. True to its tendency of surfacing at times of upheaval, Ezsenyi's *Twelfth Night* was one of the last productions staged before a drastic change in the National Theatre leadership in 2004.

In contrast to the fractured *Twelfth Night* at the National Theatre, the Summer Shakespeare Festival (SSF) production delves deeply into the question of gender and its unstable performance, venturing beyond the boundaries of the status quo to question postcommunist assumptions of gender normativity. This production takes full advantage of spatial allowances the Festival provides. Unlike the highly structured and institutionalized space of the National Theatre, the SSF is a relaxed affair staged in a Renaissance courtyard of a palace within the Prague Castle complex. The productions traditionally pass for high profile 'summer entertainment' during the regular theatre 'off season', wherein audiences can relax watching high art performed by some of the most celebrated stars of the Czech theatre that they might otherwise glimpse on the film screen. Viktor Polesný's 2005 *Twelfth Night* seemed to follow suit with a cast of high-profile actors and an interpretation of the play that most reviewers found 'entertaining'. And yet, under the mantle of light entertainment, Polesný's production managed to present coherent plot development and distinctive presentation of consistent individual characters, including those usually considered minor, to take full advantage of the play's inquiry into gender performance, sexual desire and cultural inclusivity.

In this context, the performance of masculinity is one of the central foci of this production, presenting every male character as a unique showcase of various masculine traits. Whereas most productions, for instance, sideline the lovesick Sir Aguecheek for comic relief, Polesný's uniquely foregrounds the serious implications of his courtship of Olivia and highlights his devoted attention to Sir Toby. His ineffective adoration, combined with the realization of Olivia's utter indifference to both his suit and his mere existence, results in a portrayal of explicit – and laughable – masculine failure, further underscored by frequent non-verbal comparisons with the ambiguously-gendered Cesario, who nevertheless enjoys greater success in securing Olivia's interest. These comparisons seamlessly crystallize into Cesario's and Aguecheek's provoked sword-fight in Act III, which makes visible the implicit competition for the embodiment of manhood, exemplified by the two characters' ineptitude with their swords. The two who wield swords effectively and resolutely, Antonio and Sebastian, appear in this performance as a loyal, unquestionably masculine pair. Antonio leads as an older, experienced, wise and weather-worn seaman, fully aware of the implications of his actions – wielding all of Shakespeare's original lines, traditionally cut in most performances – in proposing to wager his life for Sebastian's proximity (see Figure 5.5). Sebastian's erratic behaviour seems rooted in grief for his supposedly drowned twin sister as well as in confusion about the intensity of his connection to Antonio. His positive reaction to Olivia's insistent wooing is reminiscent of a child awakening in a sweet shop rather than a self-propelled quest for a partner. In short, unlike other performances of *Twelfth Night*, Polesný's version invites the audience to contemplate

Figure 5.5 Jan Novotný as Antonio and Jiří Racek as Sebastian. *Večer Tříkrálový* (*Twelfth Night*), dir. Viktor Polesný, SSF.

the possibility of a spectrum of masculinities that stretch beyond the twentieth-century equation of masculinity and exclusive heterosexuality. Antonio and Sebastian, after all, are simultaneously the most masculine and accomplished of the *Twelfth Night* roster; they are *also* explicitly affectionate with each other.

In addition to not shying away from a non-traditional portrayal of Antonio and Sebastian, this production prods further into the assumed centrality of masculinity to human existence. At the close of the play, twentieth-century performances of *Twelfth Night* traditionally make Olivia gleefully accept Sebastian as a rightful substitute for – and decided masculine improvement over – the reluctant Cesario; as a case in point, we have seen the explicit joyful consummation of Sebastian and Olivia in the National Theatre. This happy resolution of the dilemma of a woman being in love with a cross-dressed woman is also traditionally accompanied by non-verbal evidence of Olivia's embarrassment at her attachment to Cesario, finally unveiled as a 'maid'. In Polesný's production, Olivia's affection remains with Cesario, whom she had passionately kissed earlier in the play, marking one of the first female same-sex kisses

Figure 5.6 Jitka Schneiderová as Cesario and Linda Rybová as Olivia. *Večer Tříkrálový* (*Twelfth Night*), dir. Viktor Polesný, SSF.

in Shakespeare (see Figure 5.6), and she shows considerable distress at having realized that she has married a substitute. In this sense, this production radically departs from the tradition of *Twelfth Night* interpretation as well as traditional understanding of gender identity and masculinity. After all, it is clear that, in this version, the most eligible bachelorette remains ambivalent about the fact that she has been positively tricked into heteronormativity and, despite of what anyone may say about the similarity of near-identical twins, attaching herself to a virtual stranger.[28]

As much as the production breaks out of the mould of asserting traditional models of masculinity as the exclusive template of 'normal' male behaviour, it simultaneously carefully contextualizes the stage action in frameworks that detract from the urgency of the production's cultural implications. The minimalist outdoor stage (see Figure 5.7), used as a backdrop to the entire play, continuously superimposes an aura of the 'unusual' and frivolous. Constantly moved and manipulated (the seams of the backdrop reveal where the

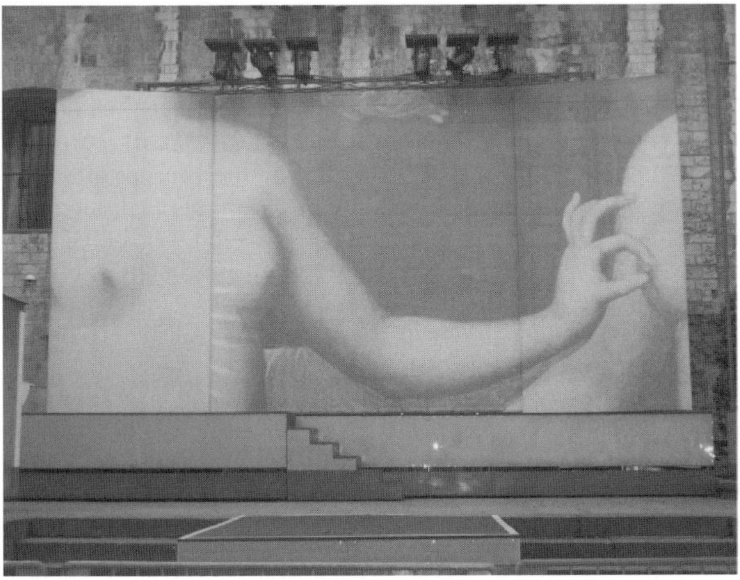

Figure 5.7 Scene backdrop to *Večer Tříkrálový* (*Twelfth Night*), dir. Viktor Polesný.

two smaller wings swing forward), the set of the play further evokes an atmosphere of playfulness and renaissance eroticism, presenting the action as delicate yet provocative sexual stimulation, rather than critical discourse.

The sense of erotic playfulness and general entertainment is echoed by the production's reviewers. Despite the director's and dramaturge's repeated public claims (in varied interviews) that their *Twelfth Night* was a 'romantic comedy about love with a bitter melancholic undercurrent and a problematic happy end' (rh, 2005: 11), as if by mutual consent, none of the performance reviews comment on any of the sexual relationships that challenge the status quo. The reviews generally connect the subtitle 'what you will' with the various forms of love presented in the play. Some, like Saša Hrboticky's review, exemplary of this approach, even capitalize on the argument of the heterogeneity of erotic expression in this play, stating that the entire production was built on 'narratives of various forms of love, ranging from self-deification to selfless self-sacrifice, and, as is typical for the author [Shakespeare], this is enriched by multi-layered interpretive possibilities' (Hrboticky, 2005: 7). Nevertheless, the variety of 'love', as described by the reviewers, amounts to the fairly homogeneous complicated and largely unsuccessful heterosexual courtships, led by the connection of Viola to Orsino and the unhappy attraction of Olivia to Cesario. While varied interpretations of the unsuccessful courtship of Olivia by Sir Augcheek and Malvolio seems integral to most of the reviews, few mentioned Sebastian as a 'happy' substitute for Cesario and *none* admitted Antonio's existence.

Instead of reading the main characters' ambivalent relationship to their eventual marriage partners as a venue of problematizing the play's generally-accepted happy heterosexual resolution, acknowledging the 'problematic happy-end' foregrounded by the production's director and dramaturge, reviewers who commented on the production's ending invariably expressed disappointment with what they perceived as shortcomings in direction or even individual acting skills. Simona Polcarová, for instance, commented on the lack of space that Orsino's character negotiates in this particular production, which results in the disappearance of his 'confused emotions toward the disguised Viola' which in turn causes the unfortunate result that 'the two couples of lovers at the end ... therefore have the unexpected effect of a fist in the eye' (2005: 15). Similarly, Josef Mlejnek was critical of Viola's diminished attraction to Orsino, which seemed lost in light of her 'confusion over Olivia's infatuation' as well as of Orsino's seeming lack of interest in Viola at the end, whom he accepts as a suitable left-over that saves his position (Mlejnek, 2005: 3–4). Overall, the reviewers, either unaware of the production's alternative interpretive possibilities or unwilling to disturb the otherwise pleasant coherent summer entertainment, underscored the view that Polesny's *Twelfth Night* is 'a light comedy without an original interpretation, which does not seek to challenge the specific [Summer Shakespeare Festival] audience' (jš, 2005: 1).

The response of the reviewers attests not only to the production's tentative presentation of unorthodox behaviours in Shakespeare, but also to the reviewer's own function within a cultural context in which a positive representation of non-normative masculinities in Shakespeare presents a cognitive glitch. Whereas the alternative Montmartre *Sonnets* could be culturally contained by media silence, the prominence of the SSF *Twelfth Night* – pooling the popularity of the venue, the reputation of Polesný as a director, the visibility of the production's stellar cast and the traditional media hype about new SSF productions every spring – made an outright boycott impractical. Instead of erasing uncomfortable cultural representations of Shakespeare, media silence in this case would have created controversy, voluminous publicity and inevitable cultural dialogue. In this scenario, turning a blind eye to the potentially disturbing elements of the production provided an easier path towards the reinforcement of traditional cultural normative boundaries. Not surprisingly, many of the reviews turned to issues traditionally raised in *Twelfth Night* productions, bolstered by the worn-out claim to the 'endless multiplicity of interpretive possibilities' ascribed to Shakespeare's genius without endorsing the one in question.

When we examine *Twelfth Night* performances and focus on the various renditions of masculinity they offer, it becomes clear that Shakespeare – despite the text of the play – is handled in ways that, at best, reflect larger confusions about masculinity in the postcommunist Czech culture and represent this confusion as profoundly detrimental to the future developments of the country. At worst, these productions reinforce the traditional status quo presented as a template for a viable future of Czech society. In their ambivalent stance towards homosexuality, the productions signal that the Czech culture in general, however willing to tolerate homosexuality either in the abstract or in the exaggerated representation that is to entertain, seems committed to marginalizing both representations of homosexual desire and homosexuality itself. Willing to extend homosexuals the rights necessary to pass under the watchful eye of the EU and appease its pressure for legislation of equal human rights, the culture has been so far unable to allow the issue into its field of vision as an everyday reality. In this context, the unwillingness to accept positive representations of non-normative masculinity in Shakespeare plays speaks to both the centrality of Shakespeare to an understanding of Czech national identity and the unacceptability of homosexuality as a viable form of postcommunist subjectivity.

Epilogue: Into the European Union

In *The Power of the Powerless*, written in the last decade of communist rule, Václav Havel hypothesizes about the intellectual and behavioural responses of citizen subjects dominated by a pervasive 'post-totalitarian' system. Here he outlines a widespread practice of living 'the lie' of the double-consciousness that the post-totalitarian system elicits. While most subjects of the system are fully aware of the system's fallaciousness and corruption, they nevertheless participate in its constitutive rituals to avoid detection and subsequent sanctions that would jeopardize their existential prospects. This complicity with the post-totalitarian system, Havel persuasively argues, however rooted in an understandable desire for self-preservation, nevertheless strengthens the repressive powers of the state.

To disrupt the self-perpetuating cycle of the pervasive power of the post-totalitarian system, Havel foregrounds the concept of 'living in truth' wherein individuals engage in a range of actions and behaviours that refuse to perpetuate the cycle of falsehood. Havel's 'living in truth' encompasses a range of possible behaviours, of which open opposition to the political system is but a fraction; a truthful life can be attempted via any action (or non-action) that provides an alternative to behaviours required by the system in power:

> in its most original and broadest sense, living within the truth covers a vast territory whose outer limits are vague and difficult to map, a territory full of modest expressions of human volition, the vast majority of which will remain anonymous and whose political impact will probably never be felt or described any more

> concretely than simply as a part of a social climate or mood. Most of these expressions remain elementary revolts against manipulation: *you simply straighten your backbone and live in greater dignity as an individual*. (Havel, 1985: 64–5, emphasis added)

Havel's parameters of a 'truth' likely to disrupt a post-totalitarian system are left deliberately vague to leave the field open to a variety of resistant, disruptive and independent actions and behaviours whose ethical dimensions are implied, but are not clearly articulated (Havel believes that human interiority is inherently ethical in its desire to live with integrity). This multifaceted 'truth', which presumably enables all citizens to pursue their discreetly-defined life projects within implied common ethical bounds, is the defining cornerstone of a viable society, free of state interference in private matters and integral to the human ability to live the life of dignity (ibid.: 69).

Far from remaining in the realm of academic philosophy, as I have sought to establish in Part I of this book, Havel's 'living in truth' became the organizing slogan of the Czech Velvet Revolution (and had spread, in equivalent variations, to other CEE countries at the brink of shedding the communist systems), signalling a promise of future freedom from artificial control of individual subjectivity, and a real possibility of living with ethical integrity in whatever postcommunist system would develop. And yet, as Czech citizens quickly learned, the postcommunist transition created new equivalents of state interference. Eerily reminiscent of aspects of communist state control, this new system provided rosy public rhetoric of individual and social freedoms, while undermining the very material prerequisites for lives of integrity: employment structures, social structures and the public political sphere. Despite the reconstitution of democratic frameworks in which free elections became an inherent expectation, democracy has been compromised by foreign economic and political pressure on the Czech government bodies to align the Czech socioeconomic and political systems to optimize the execution of Western interests in the country, predominantly at the expense of its inhabitants. For all public appearances, democracy materialized fully in the postcommunist era, in that citizens elected representatives from a multitude of political parties who ran on a wide variety of political, economic and social platforms. Nevertheless, despite the rhetoric – and even process – of functioning democracy, Czech citizens watched in dismay as those representatives, once elected, were cornered by external political powers to create legislative frameworks – outlining the normative parameters of Czech postcommunist society – that reflected, instead of the mandates on which the representatives were elected, the mandated interests of the EU and the West.

The outcomes of neoliberal practices implemented in the 1990s in effect existentially restricted the majority of the Czech population; it is not surprising, then, that Czech citizens, though inevitably participating in the rituals required by the external political frameworks for the necessary accession to the EU, continued to search for alternative parameters of 'truthful' private subjectivity that would provide a degree of integrity in the postcommunist neoliberal system. Considering the tradition of turning to the cultural sphere at times of perceived threat to national integrity to draw on a repository of essential cultural values where central tensions can be addressed, it could be expected that the cultural sphere would once again closely reflect the incipient political tensions. As this volume sought to demonstrate, within this cultural sphere, Shakespeare has figured prominently, providing a ready traditional repository of presumed transcendental cultural values that could be tapped for exploration in the uncertain transitory period. Though not unified in interpretation, as the public tensions between prominent Shakespeare translators explored in Chapter 3 exemplify, Shakespeare has provided both a site for cultural debate about suitable postcommunist subjectivity and a platform of resistance against perceived cultural threats wrapped in the unprecedented, mandatory and non-negotiable economic and cultural requirements that were frequently inconsistent with established practices of the existing EU member states.

As the subsequent chapters sought to demonstrate, productions of Shakespeare plays have reflected ongoing tensions about the parameters of postcommunist subjectivity. Residual resistance to the legacy of communist-imposed gendered subjectivity, for instance, clashed with gender equality requirements of the EU. The wide adherence to essential concepts of gender, creatively projected into postcommunist productions of Shakespeare's plays, has signalled an attempt to recover a semblance of 'truth' within established, universalized frameworks of human subjectivity. In as much as a subjective performance of hyperfemininity, for instance, provided a meaningful site of individual opposition to the communist ideology of the socialist worker-mother, essentialized, hierarchically-conceived gender roles provide frameworks of empowerment resistant to EU rhetoric of gender equality for materially and symbolically disenfranchised postcommunist subjects.

In tandem with this, the complex way in which the pervasiveness of domestic abuse in the postcommunist Czech society is projected into concurrent productions of *The Taming of the Shrew*, explored in Chapter 4, suggests that the abuse is not a simplistic backlash of emasculated Czech men misplacing their insecurity onto the more vulnerable members of the Czech society. Similarly, the icy reception of non-normative masculinities in representative Shakespeare productions of *The Sonnets*, *The Merchant of Venice* and *Twelfth Night*, which were discussed in Chapter 5, extends beyond the engrained homophobia in the postcommunist Czech culture. Simplifying these multilayered social issues into problems of postcommunist nationhood would miss the central role that Western policies have played in the programmatic material disenfranchisement of the Czech citizens, while reinforcing the Western paternalistic rhetoric of necessary interference in the social structures of a state incapable of solving its own social politics. Rather, the adherence to an essential gender hierarchy evidences a collective framework squarely oppositional to the perceived self-interested Western manipulation of the normative parameters of Czech postcommunist subjectivity, a manipulation similar to the organization of Havel's post-totalitarian system 'in which people are ... organized in one way or another (by someone who always knows best "what people need") so that they may then allegedly be liberated' (Havel, 1985: 69). The postcommunist oppositional framework, then, merely attempts to stage an 'elementary revolt to manipulation'. straightening the collective 'backbone [to] live in greater dignity'. In this context, even violent measures of achieving 'truthful' gender relations are widely perceived as the unfortunate, profoundly regrettable, but sometimes necessary byproducts of nationalistically independent project of postcommunist subjectivity-building.

Even as 77.3 per cent of the 55 per cent of Czech voters who turned out in the June 2003 EU accession referendum expressed their assent to join the European Union,[1] the Czech government struggled to balance the Czech nation above the turmoil of Western cultures divided over the imminent war in Iraq and subsequent tensions in world politics. The Czech Republic and Poland – two of the nations poised at the brink of joining the EU in the first vanguard of passing from the East into the West – were the only CEE countries to seek to join the 'coalition' Western forces that entered Iraq in the campaign to depose Iraqi president Saddam Hussein headed by the USA. The membership in this coalition was far from easy. Like most of the world, the Czech Republic was, on the one hand, under pressure from the US and British administrations, eager to use the involvement of CEE countries in their military reassertion of imperial might as a validation of their actions. On the other hand, the Czech Republic had to answer for its decisions to its future colleagues within the EU. Most significantly, the Czechs had to answer to Germany and France, who were most vocally opposed – on ethical, economic and military grounds – to the US/British venture. This ideological, political and economic split between these Western superpowers created a shattering wedge in the perception of the West as a unified cultural, economic and political entity. Even if it had previously been hard to categorize the 'West' as a homogenous conglomerate of states ruled by an identical set of cultural values, this rift signalled an undeniable end to the 'common culture' in which the Czech lands grew to their full cultural flowering from the fourteenth to the early twentieth centuries, and towards which the Czech culture has been striving in the process of shedding its communist past.

This fissure was particularly damaging to the illusion of a unified 'Westernism' because of the rhetorics employed by the main players, particularly in their discussion of the 'ethics' of preemptive war and responsibility for the political development of formally non-Western countries. On the far side of the Atlantic, the US administration more or less articulately argued for the essential responsibility of the West – consisting of privileged, 'developed' and economically powerful countries – to protect the safety and consistency of the West and to bring the values and political structures of (Western) democracy to states deprived of this civilizational privilege. The US administration repeatedly cited the failure and selfishness of the European nations – particularly Germany and France, whose support would have been crucial for the validation of the project – to uphold this responsibility, which was central to the Western essence. The more immediate neighbours of the Czech Republic, in contrast, underscored the ideological and factual lapses in the US case for a preemptive war against Iraq, often citing underlying US economic 'interest' in the Middle East as the main reason for its desire to control Iraq.

Sandwiched between these two camps, which represented very different models of Westernism, the Czech administration took the opportunity to harness this opposition to lever some efficacy for the Czech Republic on the international stage. Plagued by the decisive lack of popular support for what was perceived as 'Bush's selfish War on Iraq', the Czech administration orchestrated a plan of action worthy of Švejkian diplomacy. On the one hand, the Czech administration pledged support for the invading coalition of US and British armies by sending a small military unit expert in the detection of chemical and biological weapons, later joined by a Czech military hospital. On the other hand, the Czech administration pledged to the Czech populace and its neighbours that no Czech military personnel would become directly involved in the invasion and that Czech units would remain exclusively on Kuwaiti territory. The naiveté of the Czech administration was exposed within days, when the Czech contingent was more or less forced to cross into Iraq in response to a desperate US plea to protect coalition units who may have been targetted for chemical and biological attacks (a claim that was quickly revealed to be false). This schizophrenic, contradictory politicking move on the part of the Czech government made the shakiness of both the Czech position in the Western cultural arena and the fragile – if at all coherent – identification as a cohesive member of the Western bloc conveniently visible. On the eve of joining the EU, this anecdote suggests that Czech cultural confidence in its own identity, society and essence was conflicted.

In this turbulent cultural context, Shakespeare productions and scholarship continue to reflect, resist, further and continue many of the current political debates. Particularly at times of low cultural confidence – stemming not only from uncertain international standing but also from profound shifts within perceived cultural essence, especially when relating to questions of individual rights, responsibilities, citizenship and identity – Shakespeare as a perceived universal anchor of humanity becomes useful once again as an important site of cultural debate. Even the fractured faces of Shakespeare, produced by scholars and theatre professionals whose work is fuelled by varied artistic visions, different perceptions of postcommunist reality and unequal budgets continue to revolve around an assumption of universality of Shakespeare's works that will – no matter what the artistic approach – create a parallel between the timeless centre of Western identity and Czech artistic essence.

Ten years after its induction, the Summer Shakespeare Festival (SSF) has become an artistic fixture in Prague, slowly spreading its tentacles to an increasing number of major Czech and Slovak cities. Even though it is no longer endorsed by the President of the Czech Republic – now that the quintessential artist and humanist Václav Havel has been replaced with the conservative, though now self-professedly reformed early proponent of shock therapy, economist Václav Klaus – the festival draws multitudinous audiences, selling out most performances far in advance. Proportionately, the productions become more complex each year, drawing together more famous (and more expensive) directors and actors. As much as a cultural spectacle, the SSF has become a national spectacle, showcasing some of the largest Czech artistic names right next to that of Shakespeare.

In contrast to the pronounced success of the SSF, theatres in general have continued to struggle for new artistic identities that would attract committed audiences and to develop new ways of financial survival. The alternative theatre Komedie, a venue that belongs to a small group of theatres sponsored by the city of Prague, lost its funding. The governance was transferred to a new artistic director who brought with him a new cast of actors, producers and theatre artists, thereby completely misplacing Komedie's famed mission and execution of theatrical texts. However, in a typical Czech move of paradoxical unpredictability, while the state exercised its power in disbanding the culturally unruly Komedie team, its most prominent director, Michal Dočekal, won the artistic competition for the artistic directorship of the floundering National Theatre. In one sweeping move, the city of Prague simultaneously expressed its ambivalence about the perhaps most alternative of its stages and enabled its leader, one of the youngest to be appointed to a position of decisive power within the institutional behemoth of the National Theatre, to spearhead the struggling national stage towards new horizons.

As much as Shakespeare has been the centrepiece of Czech cultural identity, the number of Shakespeare productions seems to be decreasing in the aftermath of the military engagement in Iraq and Afghanistan, which highlighted the pervasive inability of CEE countries to harness postcommunist efficacy in the face of the much larger international political competitions. In May 2004, four years after this project was begun, most Prague theatres provided about one Shakespeare night for its audiences, totalling 21 performances in the month, less than one a day. Though perhaps a significant number in comparison to the life of the theatre in other cities of comparable size, it is a marked decrease from May 2000, when there were several Shakespeare productions at several venues competing for the audience's favour every evening. The most visible difference is in the number of Shakespeare performances at the Prague Globe: whereas in May 2000 the Globe performed nightly, often adding matinees and countless performances and Shakespeare programmes for schools in the mornings, in May 2004 it scheduled a mere seven performances in the entire month. The space opened up by the partial withdrawal of Shakespeare plays seems to have been quickly filled with Czech (both historical and new) as well as foreign (overwhelmingly Western) plays.

This book ends on the eve of incorporation of a number of former Eastern European countries into the new 'Western' Europe, even as universal Shakespeare seems to give more room to new, original Czech theatre that reflects new directions of understanding both the position of the individual within society and the position of the society as a whole on the spectrum of 'civilized' Western cultures. Inevitably, the project continues beyond the last page. The analysis of political uses of Shakespeare remains a rewarding enterprise. Considering that canonical art has been, and will most likely continue to be, in one way or another, at the centre of Czech cultural identification, artistic – and particularly theatre – productions will further mirror issues most important for audiences who choose to attend them, issues that will remain political even when they fall into the category of personal taboo.

Many questions remain even within the context of Czech Shakespeare: what shapes will Shakespeare take in response to the rapidly changing national and international political scene? Where will his works fall on the spectrum of Czech cultural consciousness? Will he be relegated, as the Czech resistance to European policies rises and Czech national identity solidifies to a comfortable degree cultural confidence, to a mere footnote on the syllabi of English literature courses? Or will the presumed transcendence of works outweigh the rising taint of corrupt and exploitive Westernism? If Shakespeare is to remain close to the beat of the Czech cultural heart, what culture(s) will his works begin to reflect? The future is unknown. On the literal eve of Czech fusion with the EU – 30 April 2004 – the primetime nationally broadcast celebration at the National Theatre showed, side by side, examples of both Czech and Western artistic achievement. Not surprisingly, the most prominent artistic entity among them was Shakespeare.

Notes

Introduction

1. In her analysis of the 'Challenges of EU Enlargement', Heather Grabbe has noted that the accession conditions move far beyond any rules or regulations that the existing members apply to themselves:

 > It is even hard to say whether all the current member states are 'ready to join,' because they have never been judged on the Copenhagen conditions. For example, Belgium's public administration, France's industrial policy, and Germany's state aids would probably be unacceptable if they were applicants. Some Commission officials claim that no current member state fully implements more than 85 percent of EU regulations. So is it fair to demand 95 percent adherence from the east Europeans? (2003: 78).

 Further, as Grabbe outlines, the situation is exacerbated by the reality that previous instances of EU enlargement were marked by slow, negotiations that extended well into the accession process, took into consideration the interests of all parties and were supported by large financial investment from existing members. Non-negotiable conditions, defined by existing members who stipulate evolving targets that are constantly moving yet need to be satisfied *prior* to accession, are a complete novelty within this context and signal a new, likely hierarchical development within the nominally egalitarian Union.

1 The 'End of History' in Central and Eastern Europe

1. Katherine Verdery echoed this observation in her analysis of the postcommunist transition, charting the ways in which both dissidence to communism and early postcommunist developments were defined by the rhetoric of returning to Europe (Katherine Verdery [1996] *What was Socialism, and What Comes Next?* Princeton University Press, 1996: 104). Numerous scholars have pointed to the links between neoliberalism and the rhetoric of democracy; see, for instance, the exemplary analyses by David Harvey in *The Brief History of Neoliberalism* (New York: Oxford University Press, 2007), particularly Chapters 1–3; Naomi Klein in *The Shock Doctrine: The Rise of Disaster Capitalism* (New York: Metropolitan Books, 2007); Henry A. Giroux in *The Terror of Neoliberalism: Authoritarianism and the Eclipse of Democracy* (Boulder: Paradigm, 2004); or Lisa Duggan in *The Twilight*

of Equality: Neoliberalism, Cultural Politics, and the Attack on Democracy (Boston: Beacon Press, 2003), particularly Chapter 1.

2. As historian Robin Okey points out, while the first Marshall Plan was equal to 1% of the US gross national product (GNP), the financial commitment to CEE after 1989 was equal only to about 0.02% of the GNP of the Western members of the Organization of the Economic Cooperation and Development (OECD). Though in 1949 the Marshall Plan contributed 11.5% of France's GNP, postcommunist Hungary – one of the primary early recipients of Western aid – received only 0.5% of its annual budget; incidentally, the cost of 'structural reforms' imposed by the EU in most CEE countries will swallow up about 6% of the GNP of each applicant country (Robin Okey [2004] *The Demise of Communist East Europe: 1989 in Context*. Oxford University Press: 181–2). Of the allocated transition funds, only a small proportion consisted of grants (researchers assign between 10 and 20% in this category) as compared to the 80–90% of the original Marshall Plan.
3. Similar argument has been made by other scholars of the postcommunist transition: William Wallace ([2003] 'Does the EU have an *Ostpolitik*?', in Anatol Lieven and Dmitri Trenin (eds) *Ambivalent Neighbors: The EU, NATO and the Price of Membership*. Washington: Carnegie Endowment for International Peace: 51), Zaneta Ozolina ([2003] 'The EU and the Baltic States', in ibid.: 214), Christopher Bobinski ('Polish Illusions and Reality', in ibid.: 243), Vladimir Baranovsky ('Russian Views on NATO and the EU' in ibid.: 271) and Anatol Lieven ('Conclusions: The Pangs of Disappointed Love? A Divided West and its Multiple Peripheries', in ibid.: 304).
4. Furthermore, instead of ensuring that all citizens possess meaningfully sufficient material means that predicate basic participation in a democratic structure, neoliberal parlance equates 'freedom' with freedom of economic enterprise. Always aimed at increasing the material profit of those initially able to invest substantially, neoliberalism inevitably redistributes available capital towards the wealthy echelons of society, inescapably at the expense of the general society and its economic security. Through the weakening of the state support system, general individual freedom thus decreases, while the freedom of those holding the reins of resources grows further. To protect this framework, neoliberalism tends to limit democracy to produce results that would buttress the status quo. As David Harvey has articulately argued in *The Brief History of Neoliberalism*: 'While individuals are supposedly free to choose, they are not supposed to choose to construct strong collective institutions (such as trade unions) as opposed to weak voluntary associations (like charitable organizations). They most certainly should not choose to associate to create political parties with the aim of forcing the state to intervene in or eliminate the market. To guard against their greatest fears ... the neoliberals have to put strong limits on democratic governance, relying instead upon undemocratic institutions (such as the Federal Reserve or the IMF) to make key decisions. This creates the paradox of intense state interventions and government by elites and "experts" in a world where the state is not supposed to be interventionist' (2007: 69).

In such a context, Harvey argues, democracy becomes a 'luxury' of which the current citizen subjects are unworthy (ibid.: 66).

5. For an exhaustive survey of the origin and impact of neoliberal practices around the globe, see, for instance, Harvey (2007) or Klein (2007).
6. In addition to profiting from traditional colonial venues, neoliberalism expands the frontier of capitalist profit to the crucial sphere of publicly-held assets. Through the persistent insistence on privatization of national enterprises, neoliberalism assists in the redistribution of existing wealth from democratically-constituted public hands to the hands of wealthy private individuals and corporations. Commenting on the neo-imperial attributes of neoliberal practices, Naomi Klein argues: 'Smith's colonists earned their record profits by seizing what he described as "waste lands" for "but a trifle," today's multinationals see government programs, public assets and everything that is not for sale as terrain to be conquered and seized – the post office, national parks, schools, social security, disaster relief and anything else that is publicly administered. ... the state acts as the colonial frontier, which corporate conquistadors pillage with the same ruthless determination and energy as their predecessors showed when they hauled home the gold and silver of the Andes' (2007: 242). Rather than accepting assistance, then, countries that agree to employ neoliberal practices under the tutelage of Western experts have generally made themselves subject to neo-imperial exploitation that, contrary to the official humanitarian clauses of such assistance, further impoverishes their economies and people, while tying up their independence in growing and unpredictable debt-service obligations.
7. Aside from the selective Western adherence to the international codex of human rights, which tends to undermine meaningful attempts at human rights enforcement in non-Western territories, it is difficult to reconcile the substantial equality the codex assumes at its base with the systematic inequality neoliberalism has inevitably produced wherever it has been implemented. Despite its core rhetoric of protection of the right of the individual, neoliberalism tends to *produce* – rather than eradicate – human rights violations. Naomi Klein again has attested, 'The casual exclusion of tens of millions of people by the free-market ideologues has reproduced frighteningly similar explosive conditions [to Hitler-led fascism]: proud populations that perceive themselves as humiliated by foreign forces, looking to regain their national pride by targeting the most vulnerable in their midst' (Klein, 2007: 450). Apart from this regressive and explosive nationalism, noted throughout lands where neoliberal practices were implemented, increased competition for resources creates its own existential tensions that, in their extreme forms, result in regional sparks of violent struggle for material survival.

   Even philosophically, in spite of its rhetoric of individual freedom, neoliberalism in practice has proven itself profoundly suspicious of 'excessive' political and economic autonomy of the individual. While meticulously policing the property rights of the wealthy, neoliberalism has paired with the conservative movement in an intense backlash against social and civil freedoms of traditionally disenfranchised sectors of the populace.

While neoliberalism employs the façade of individual freedoms to harness popular consent for its disenfranchising policies that assists the accumulation of capital in the hands of the wealthy, neoconservatism takes stock of the usually dismal material outcome of neoliberal practices to blame 'excessive' individual freedom, usually as it is embedded in civil rights. Rather than unveiling the impact of neoliberal practices, neoconservatices decry the provisions of 'special groups (blacks, women, environmentalists)' as detrimental to the collective well-being (Harvey, 2007: 50).

2 Shakespeares of the Postcommunist World

1. For more detailed analysis of the process of incorporating Shakespeare into the early communist canon, see Ostrovsky's 'Shakespeare as a Founding Father of Socialist Realism: The Soviet Affair with Shakespeare' in *Shakespeare in the Worlds of Communism and Socialism*. University of Toronto Press, 2006: 56–83.
2. For an articulate, well-researched, current account of Shakespeare scholarship produced during communism, see Irena Makaryk and Joseph Price's *Shakespeare in the Worlds of Communism and Socialism* (University of Toronto Press, 2006), particularly the essays in sections 1–3.
3. Sokolyansky is hardly a lone voice on the Russian Shakespearean scene. The main thrust of an international 'Shakespeare Readings' conference in June 2002 in Vladimir was very similar.
4. This reference is so pervasive (framing every chapter as the organizing extended metaphor) that a page-number reference would be meaningless.
5. Socialist realism favoured the 'people' and its struggle for ideal communist society. At the centre could usually be found a 'socialist hero' who functioned as a transformative and motivational force to those around him in building a proper communist society. Arkady Ostrovsky, a contributor to *Shakespeare in the Worlds of Socialism and Communism*, in the context of discussing socialist-realist interpretations of Shakespeare, summarizes the 'distinguishing marks of Socialist Realism' as 'clarity, truth-to-life, moralism, hard-line didactism and a striving for clear-cut simplicity. Adjectives like elusive, oblique, fluid, rare, sensitive, mutable, airy melting were no longer part of the critical vocabulary' (Ostrovsky, 2006: 62).
6. This insistence on stepping outside the systematic falsehood of totalitarian communism is hauntingly anticipatory, by some 20 years, of Václav Havel's *Power of the Powerless* that famously describes the everyday deliberately falsified existence of communist subjects to argue for collective disruptions of the systems of 'living the lie' by 'living in truth'.
7. The cornerstone for the National Theatre was laid in 1868, though the building was not finished for nearly another decade, due to a fire that destroyed the nearly finished first building. The building and rebuilding of the theatre is legendary as a national endeavour, wherein all Czech patriots, including the poor, were said to contribute their meagre savings to this visible symbol of Czech nationalistic spirit.

8. Jakub Malý, Neruda's contemporary, similarly recognized the importance of Shakespeare in the space of Czech national development and theatre, for his mere presence evoked 'not only wider education, but the entire pure humanity' (Malý [1873], 'Shakespear a jeho díla', Prague: Muzeum Království Českého: 1). Though critical of German appropriations of Shakespeare as bardolatrous (ibid.: 54–6), he celebrated Czech appropriations of the playwright and their implicitly correct understanding of Shakespeare as a pinnacle of proper masculinity: 'a handsome man, of brave and grave stature, noble and direct behavior which brought him the friendship of all he had met' (ibid.: 39). The competition with German-language scholarship, thought and politics in both Neruda's and Malý's treatises is far from implicit; on the contrary, it is central to the literary and theatrical discourse of the Czech rebirth.
9. Bedřich Smetana (1833–87) was a prominent composer of the nineteenth century. Many of his works were based on Czech nationalistic themes.
10. Pečírka also recalls the 1864 Shakespeare tercentenary as a 'national manifestation' and a 'celebration of our very nation, declaring such powerful admiration for the poetry of the English bard ... the third centenary of Shakespeare's birth...was the most magnificent celebration of our own art, ... [so that] the only event of the nineteenth century which equals it in political importance and as an indication of the cultural level of the people was the laying of the foundation stone of our National Theatre' ('Shakespeare and the Graphic Arts', in Zdeněk Stříbrný (ed.) *The Charles University on Shakespeare*. Prague: Univerzita Karlova, 1966: 90, 91).
11. This production has been described by many scholars, for instance, Martin Hilský (1994: 133 ff) and Stříbrný.
12. This protest march was annually initiated to commemorate the anniversary of an execution of nine university students by the Nazi regime to silence intellectual opposition to the fascist occupation, a move that was further underscored by the subsequent closing of Czech universities for the duration of the Second World War. This annual event, while inevitably commemorating historical events, took on the far more important political dimension of parallel resistance to intellectual control exerted by the communist system. 1989 marked the 50th anniversary of the event which, together with the already intensifying frequency of public acts of regime defiance, brought together an unprecedented number of protesters whose clash with the police proved the final feather that mobilized the nation. A rumour of a student death at the hands of the riot police, later proved to be false, triggered national outrage manifested in the outpouring of hundreds of thousands of citizens onto the streets in mass protest followed by a general strike, and rebellion of the media, and even some of the armed forces that refused to 'pacify' fellow citizens.
13. Except for the brief period between the World Wars, the Czech lands were subject to imperial control not only for the four decades of Moscow-oriented communist politics, immediately preceded by German fascist occupation, but for four previous centuries of sometimes violent colonial repression by the Austro-Hungarian Empire.

14. Wedel attests that this refusal was directly related to the Czech government's articulation of a hierarchical division of the world and an explicit insistence on the country's belonging to the Western bloc, rather than an outright articulation of a resistance to neoliberal practices: 'If the Third World was defined as a region in which governments received development assistance, and if accepting aid signalled that we're not yet part of the "West" and meant remaining in a supplicant position, then the Czech government's stance was a way of defining the nation as already part of the West, refusing risks of any affronts to national pride.' Defying the Western neoliberal blueprint for postcommunist economic organization, the coordinator of aid for postcommunist Czechoslovakia argued that 'Many Czechs now proudly believe that Westerners have little to teach us, to show us, to advise ... The attitude has been essentially that we don't need the money' (Nanine R. Wedel [1998] *Collision and Collusion: The Strange Case of Western Aid to Eastern Europe, 1989-1998.* New York: St Martin's Press: 40–1).
15. The Prague Theatre Institute only began keeping central and comprehensive records and statistics for all Czech stages in 1997, but even these are quite revealing of the firm place Shakespeare has claimed on the Czech theatre scene. In the first season for which the statistics are available, 1997–8, Shakespeare was the most performed dramatic author with 614 performances of his plays country-wide, a figure which was nearly twice as high as the runner-up popular Czech trio of Smoljak/Cimrman/Svěrák, whose plays were staged 322 times. Shakespeare also led in the number of premiers of his plays (19), whereas Chekhov in the second place claimed 10. Shakespeare reportedly sold the most tickets, outshining not only all other dramatic authors but also all composers of opera, ballet, musicals and vaudevilles, with a total of 196,055 people in attendance. His plays were performed in 27 different theatres, whereas the runner-up Chekhov claimed only 14. That year, the most performed plays were *Romeo and Juliet* and *The Taming of the Shrew*, with 125 and 90 documented performances respectively, both in the top 10 most performed plays in the season: Ondřej Černý and Vencl Vít (eds) (1999) *Divadlo v České Republice 97–98*. Prague: Divadelní Ústav: 577–89. As the statistics show, Shakespeare's popularity only grew in the subsequent years: in 1998–9, Shakespeare was unvanquished at 203,347 sold tickets, a number that rose by about another 50,000 by the following year (263,050). Theatres not only continued successful runs of existing productions, but sought new plays at breakneck speed: in 1998–9 theatres staged 22 new productions, more than the premiers of the canonical authors in second and third places, Dvořák and Molière (11 and 10 openings respectively), a number that increased to 25 by the following year. In the 1999–2000 season, Shakespeare began dominating a new category of the musical, where an adaptation of *Hamlet* by Janek Ledecký claimed the first place over all other musicals that year with a total of 78,807 tickets sold, the most not only for a musical, but also for any privately owned venue: Ondřej Černý and Vencl Vít (eds) (2000) *Divadlo v České Republice 98–99.*

Prague: Divadelní Ústav: 623–53; and (2001) *Divadlo v České Republice 99–00*. Prague: Divadelní Ústav: 642–91. Shakespeare's dominance of the Czech stages had not diminished by the end of the 2002–3 season, the last one for which published statistics were available at the time of writing (summer 2008) or for the 2003–4 season, for which the Theatre Institute was kind to provide preliminary data for the purposes of this volume: the season saw 20 new Shakespeare productions in addition to 36 continuing productions from previous seasons; Shakespeare that season netted 218,421 paying audience members.

16. This information is a synopsis of data available at the official website of the festival, at www.Shakespeare.cz, personal interviews with the director of Agentura Schok, Michal Rychlý and programmes published annually for the festival.

3 Translation Wars: Redefining Shakespeare in the Postcommunist Czech Republic

1. This attack, it must be noted, is on the surface directed against the translators (in general) who 'simplify' Shakespeare's language in order to please the consumerist mass audience. Hilský's name came up only once. And yet, since all of the examples of grievous 'inaccuracies' and 'vulgarities' come from Hilský's translation of *Hamlet* and *The Comedy of Errors*, the target of the initial missive is unmistakable. Subsequent to Hilský's public response to the challenge, Hodek's vestiges of generality fall away, and the engagement with Hilský becomes official.
2. The full citation from Cicero's Book II of Oratory is as follows:

> Now feelings are won over by a man's merit, achievements, or reputable life, qualifications easier to embellish, if only they are real, than to fabricate where non-existent. But attributes useful in an advocate are a mild tone, a countenance expressive of modesty, gentle language, and the faculty of seeming to be dealing reluctantly and under compulsion with something you are really anxious to prove. It is very helpful to display the tokens of good nature, kindness, calmness, loyalty, and a disposition that is pleasing and not grasping or covetous, and all the qualities belonging to men who are upright, unassuming, and not given to haste, stubbornness, strife or harshness, are powerful in winning goodwill, while the want of them estranges it from such as do not possess them; accordingly the very opposites of these qualities must be ascribed to our opponents.... the speakers are made to appear upright, well-bred, and virtuous men.

Cicero (1990) 'Of Oratory', in Patricia Bizzell and Bruce Herzberg (eds) *The Rhetorical Tradition: Readings from Classical Times to The Present*. Translated by E. W. Sutton and H. Rackham. Boston: Bedford Books of St Martin's Press: 240.

I am grateful to Linda Shenk for steering me toward the relevant passages.

3. For detailed discussion of the Revival, please see Chapter 1.
4. The example that Sloupová provides is not only of dissecting and explaining *Hamlet* so that a lay audience can enjoy the play, but it makes visible the intersection between the text and reality or, rather, between Shakespeare and reality. In this instance, Shakespeare, literally, can be a lifesaver, providing a literary argument for the preservation of human life.
5. Though it is questionable whether Hilský meant to claim to be an oracle for Shakespeare, one of the interviewers decided to take the comment at its face value and put it as the inch-thick heading for her cover-page interview-article in *Týdeník Rozhlas* [*Radio Weekly*], at any rate making Hilský into Shakespeare's spokesperson. *Original wording:* 'Někdy mám pocit, že zastupuji samotného Shakespeara.'

4 Katharina 'Humanized': Abusing the 'Shrew' on Prague Stages

1. Rather than using the more common referent for the play, *The Shrew*, which implicitly assumes that there is indeed a shrewish character in the play, I choose the less used referent, *Taming*, since the 'taming' of Katharina is always indisputably present.
2. For a discussion of this categorization, see for instance Felicia Hardison Londré (1996) 'Confronting Shakespeare's "Political Incorrectness" in Production: Contemporary American Audiences and the New 'Problem Plays', *On-stage Studies*, 19, 67–8.
3. Katherine Verdery, for instance, provides an astute analysis of the contradictory ways in which the conflicting communist ideology of equality in difference played out; see Chapter 3 of *What Was Socialism and What Comes Next* (1996), particularly pp. 64–9.
4. After 2001, the artistic team at Komedie was dispersed among a number of theatres around Prague. Since the theatre space itself is financially supported by the city of Prague, the theatre itself did not perish. Branches of city government responsible for art and culture found a new director and artistic team to replace Dočekal and his crew.
5. It is interesting – and significant – to note that the opinions expressed by the actors on this occasion seemed to be profoundly influenced both by the roles they were embodying on stage and, perhaps more profoundly, by the immediate sociopolitical context in which the Czech government was negotiating the forceful conditions of the EU. In subsequent informal interviews, conducted after 2005, both Matásek and Štréblová presented far more moderate views on gender relations that were critical of abuse of all kinds.
6. As convincing as this argument may seem, it is useful to recall that, at least in the Renaissance framework, a complete silence on Katharina's part would be less submissive than submissive speech. Drawing on evidence from early modern domestic manuals and literary examples, Gwynne Kennedy has shown that a wife's proper behaviour is to be mostly silent, but not

completely silent, as complete silence suggests defiance. Once a woman has reflected on her own imperfections, which is exactly what Katharina does in the final monologue, the woman is to affirm her husband while refraining from excessive expressions of her thoughts. The wife's own consciousness becomes moulded to the needs and worth of her husband, since, as Kennedy claims, 'patriarchy activates conscience as an internal self-censoring aspect of subjectivity that can undercut and check even private feelings of worth': Gwynne Kennedy (1991) 'Lessons of the "Schoole of Wisedome"', in Carole Levin and Karen Robertson (eds) *Sexuality and Politics in Renaissance Drama*. New York: Mellen Lewisnton: 114–16.

7. The production thus presented a new dimension, wherein on a small scale it re-enacted the somewhat tense and competitive relationship between the two countries after the peaceful but far from amiable division of Czechoslovakia in 1991. From this perspective, it would be easy to see a Czech Petruchio's taming of a Slovak Katharina as a form of Czech bravado and humiliation of its once close Slovak partners. The play, however, is far too complicated and nuanced to communicate such a simplistic picture. Unfortunately, the scope of this book does not allow for a more detailed consideration of this aspect of this multi-dimensional spectacle.

5 Politics of Desire: Postcommunist Czech Shakespeare and Non-normative Masculinity

1. See, for instance, Lisa Duggan, David Harvery or Henry A. Giroux.
2. The first known legislation of homosexual behaviour, in *Constitutio Criminalis Carolina* (1532), specified a punishment by burning at the stake. Like other European laws pertaining to human sexuality, it allotted death for 'sodomy', which embraced all kinds of 'deviant' sexual behaviours including bestiality and masturbation. The first version of legislation on homosexuality in force in the entire Austro-Hungarian Empire came in *Constitutio Criminalis Theresiana* (1768). Article 74 specified beheading followed by a burning of the body for homosexual behaviour, reducible to long-term imprisonment for offenders who were too young, mentally handicapped or who failed to 'complete', i.e. ejaculate. Other crimes covered by the chapter were incest, necrophilia, adultery, copulating with non-Christians and animals, as well as 'unnatural' heterosexual behaviour. The death sentence was revoked in the 1787 *Constitutio Criminalis Josephina*, when Joseph II substituted imprisonment terms, forced labour or flogging for homosexual offences, depending on the gravity of circumstances. The final Austrian law to influence Czech territories was *Allgemeine Bundesgesetz* passed in 1852, which remained in force in the Czech lands until 1950 and in Austria until 1975.
3. Jelínek himself points out numerous times that the book is the first in its field and on the topic, other than a number of medical articles that tend to treat the issue superfcially and inacurately: František Jelínek (1924) *Homosexualita ve světě vědy*. Prague: Obelisk: 46.

4. Bisexuality, which is often a stumbling block in similar debates, he labels as '*duševní hermafroditismus*', i.e. psychic hermaphroditism.
5. See Pavlína Janošová (2002) *Homosexualita v České Republice*. Prague: Karolina: 48.
6. France (1985), Denmark and Ireland (1989), the Netherlands (1992), Norway (1993) and Sweden (1994). See, for instance, ibid.: 51, 62.
7. Information about this time period is available from a variety of sources. See, for instance, Fanel, Janošová, Hromada or the website of Gay Iniciativa at www.gayiniciativa.cz.
8. At the time of the publication of her monograph in 2002, she believed that Czech toleration for homosexuality was 'by far not satisfactory', pointing to research results showing that, in 1994, homosexually-oriented persons were as likely to remain closeted as in 1988 (Janošová, 2002: 50).[8] She suspects that this reluctance to come out also may be the reason for the very low proportion of the population that identifies itself as gay or lesbian, in contrast to Western countries. According to the latest research, only 0.4% identified as homosexual, 0.3% of women as lesbian and a further 1.4% of men and 2.0% of women were not sure (ibid.: 14). Her own research shows that about 52.2% of her respondents would vote for registered partnerships for homosexually-oriented citizens, 24.5% would certainly vote against it, while 22.1% were unsure. About 61.5% responded that homosexuals should live in partnerships, while nearly a quarter (22.3%) believed that they should live in celibacy. Significantly, Janošová found large discrepancies along the lines of age, wherein the older the respondents, the less likely they were to vote in favour of registered partnerships and would instead suggest that homosexuals live in celibacy. Though the overall figures are rather optimistic, it is necessary to note that the majority of the research pool for Janošová's research was urban university students, and so, considering the conclusions made within the research in respect to age groups, the results are likely to be significantly skewed in favour of tolerance for homosexuality and registered partnerships. Overall, tolerance for homosexuality may be far less prevalent.

   One of the very few non-medical monographs documenting queer issues in the Czech culture, from the nineteenth century to the present day is Jiří Fanel's recent *Gay Historie* (Prague: Dauphin, 2002). It was originally published chapter by chapter in SOHO Revue, the monthly magazine for and about 'homosexual citizens of the Czech Republic'. In the foreword, the author specifies that the book does not mean to 'propagate' homosexuality, but rather presents a series of facts about human history. Still, Fanel's history *is* profoundly political in its aim to uncover 'homosexuals' in world and Czech history and assert their undeniable gifts to humanity. Claiming to be writing for everyone, rather than only the queer community, Fanel provides current homosexuals with cultural visibility and attempts to sway public opinion in favour of tolerating homosexuality and supporting domestic partnership legislation. The first of its kind, this ambitious project counters the sense that homosexuality is a modern invention by showing that it has its place in human history. Further, in pointing out that 'homosexuals' were not always persecuted, Fanel provides a ground for the argument that they should not be discriminated against at present.

9. *Večerník Praha,* 29 May 1998.
10. In *Svobodné Slovo,* for instance, Zdeněk Rohlíček comments on the acute accuracy of the 'fifteen-year old musical' in the current 'embarrassing discussion in the parliament about registered partnership' between 'our xenophobes, afraid of everything foreign'.
11. This is, for instance, pointed out in Milena Bártová's 'Klec bláznů nabízí únik z reality'. Bártová not only suggests that the issue of homosexuality is not all that real in current Czech society by entitling her review '*The bird cage* offers an escape from reality', but also subtitles her article 'Heterosexual Ladislav Županič excels as a homosexual character'.
12. *Večerník Praha,* 6 June 1998. Other commentary is very similar in content.
13. See, for instance, Radmila Hrdinová's 'V Karlíně se zkouší Klec Bláznů', *Právo,* 13 May 1998: 10.
14. See, for instance, Milena Pekárková's 'Muzikál Klec bláznů', *Super,* 8 June 2001: 13; Jana Soprová's 'Klec bláznů: Česká premiéra poněkud šokujícího muzikálu', *Večerník Praha,* 29 May 1998 or Zdeněk Rohlíček's 'V novém karlínském muzikálu všechno hraje ve velkém stylu', *Svobodné Slovo: Kultura,* 16 June 1998: 9.
15. Quoted in Jana Hrdinová's 'V Karlíně se zkouší Klec Bláznů'.
16. There have been only three productions of the play since 1945: Dočekal's version was first in nearly three decades, and the third production, in a regional theatre, followed suit shortly after Komedie's successful staging in 1997.
17. For instance, see Erml, 2000b: 2; or Urbanová, 1997b: 11.
18. See also Bronislav Pražan (1997) 'Divadlo Komedie: Kupec Benátský 2001', *Rozhlas,* No. 25: 4, or Erml, 2.
19. Sylva Daníčková, 'Černá dáma Sonetů', *Svobodné slovo,* 14 August 1986.
20. Such commentary often foregrounds Sonnet 66 as a particularly poignant venue of indirect expression of political resistance. Sonnet 66:

> Tired with all these, for restful death I cry,
> As to behold desert a beggar born,
> And needy nothing trimm'd in jollity,
> And purest faith unhappily forsworn,
> And gilded honour shamefully misplac'd,
> And maiden virtue rudely strumpeted,
> And right perfection wrongfully disgrac'd,
> And strength by limping sway disabled
> And art made tongue-tied by authority,
> And folly, doctor-like, controlling skill,
> And simple truth miscall'd simplicity,
> And captive good attending captain ill:
> Tir'd with all these, from these would I be gone.
> Save that, to die, I leave my love alone.

-ku, 'Shakespearovy Sonety', *Lidová Demokracie,* 19 January 1991.

21. A rephrasing of the first line of the first sonnet, 'From fairest creatures we desire increase', used by Vítězslava Šrámková as the title to her review.
22. See, for instance, Vladimír Hulec, 'Shakespearovy Sonety plynou v předvídatelném toku', *Mladá Fronta Dnes,* 2 February 1996: 18.
23. Sonnet 76:

    Why is my verse so barren of new pride,
    So far from variation or quick change?
    Why with the time do I not glance aside
    To new-found methods, and to compounds strange?
    Why write I still all one, ever the same,
    And keep invention in a noted weed,
    That every word doth almost tell my name,
    Showing their birth, and where they did proceed?
    O! know sweet love I always write of you,
    And you and love are still my argument;
    So all my best is dressing old words new,
    Spending again what is already spent:
      For as the sun is daily new and old,
      So is my love still telling what is told.

24. Between 2000 and 2005, four different theatres staged *Twelfth Night* in Prague alone; in addition to the two I discuss in this essay, *Twelfth Night* was performed at the ABC (premiere 1999) and Pidivadlo (2002) theatres. Since Czech theatres generally function on the repertory model, most of these performances ran for several consecutive seasons and, therefore, at one time or another, competed for the audience's attention. The two I chose for analysis in this limited space were particularly telling for their explicit implications in the nation-building process (Ezsenyi's *Twelfth Night* at the National Theatre) and wide popularity (Polesný's *Twelfth Night* at the SSF). In terms of exploring masculinity, I found these two performances to be representative of the four versions collectively.
25. Olivia here admits: 'I suppose him virtuous, know him noble,/Of great estate, of fresh and stainless youth,/In voices well divulged, free, learned, and valiant,/And in dimension and the shape of nature/A gracious person; but yet I cannot love him.'
26. Significantly, unlike Portia in Shakespeare's *The Merchant of Venice*, who in the notorious ring sequence establishes the primacy of her marriage to Bassanio over his loyalty to the merchant Antonio, Olivia makes no effort whatsoever to address Sebastian's proclaimed bond to Antonio. This suggests that her character does not perceive the two relationships to be in competition with each other, but rather as serving two distinct purposes.
27. A respected theatre critic and Shakespeare scholar Milan Lukeš critiques the performance for questionable 'gags' and futile attempts to make

Shakespeare accessible to the populace. These attempts, he argues, 'go so far as to overtake an identifiable contemporaneity by some futuristic vision that is an especially uncomfortable manifestation of outward modernization' (1997: 5). A compendium of theatre critics who produce weekly ratings of current performances in *Divadelní Noviny*, the premier Prague newspaper dedicated to theatre, awarded Ezsenyi's *Twelfth Night* an average of 2.55 stars of 5, on a scale on which 3 equals 'worth seeing' and 2 'at one's own peril'. Vladimír Mikulka renamed the play 'Twelfth Night, or whatever we thought of' (2001: 5), pointing out the jumbled nature of the production, while, in another review, he laments the audience's eagerness to return to see the performance, particularly its explicit sexual content (2002: 25–38, particularly 34).

28. As in most androcentric societies, homosexuality in the Czech Republic is discussed primarily in regards to male-male relationships. This does not indicate that lesbian relationships are explicitly more tolerated or respected; largely, they seem invisible. When made visible (and political), such as in Polesný's *Twelfth Night*, they are simultaneously 'sexy' and threatening of masculinity.

Epilogue: Into the European Union

1. These numbers have been widely reported and are available, for instance, in the European Gallup archives at http://www.gallup-europe.be/epm/epm_czechrepublic.htm. It is important to note that 55 per cent marks an embarrassingly low turnout for Czech voters, who vigorously partook in postcommunist elections, and the number of positive votes, 77 per cent (many of which were reported by pollsters as ambivalent), marks one of the lowest percentages among the candidate states (with only Malta, Estonia and Lithuania registering lower approval rates at 53, 67, and 67 per cent respectively). The final results reflected a much lower approval rate in the Czech Republic than the expected 82 per cent.

Bibliography

Personal interviews

Čachotský, František. Actor. Sebastian: *Twelfth Night*, Pidivadlo. 18 June 2002.

Cerqueirová, Andrea. Lesbian activist, organizer of GLBT events. 18, 26 July 2002.

Dlouhý, Michal. Actor. Petruchio: *The Taming of the Shrew*. CD 94 and SSF. 14 August 2001.

Dočekal, Michal. Director, Divadlo Komedie, Národní Divadlo (National Theatre). 23 July 2001; 14, 17 June 2002.

Duřt, Vojta. Actor. Antonio: *Twelfth Night, Úplněk*. Pidivadlo. *Úplněk*. 18 June 2002.

Frantová, Vilma. Actor. Viola, Captain: *Twelfth Night*; Anastázie: *Úplněk*. Pidivadlo. 18, 19 June 2002.

Hašek, Gustav. Actor. Fabián: *Twelfth Night*; Emil: *Úplněk*. Pidivadlo. 18 June 2002.

Helešic, Jan. Actor. Malvolio: *Twelfth Night*; Štěpán: *Úplněk*. Pidivadlo. 18 June 2002.

Hilský, Martin. Shakespeare translator; Professor of English and American Literature, Charles University. 7 August 2001; 19 June 2002.

Hlaučo, Miloslav. Director: *Twelfth Night, Úplněk*. Pidivadlo. 30 May 2002.

Hodek, Břetislav. Translator, Anglist. 7 August 2001.

Hromada, Jiří. SOHO President 1991–2001; Gay Iniciativa President, 2001–present. 17 July 2002.

Javorková, Anna. Actor. Katharina, *The Taming of the Shrew*. SSF, 2000–1. 10 August 2001.

Josek, Jiří. Translator, Professor of English and American Literature, Charles University. 28 July 2001.

Kavánková, Jana. Owner: A-Klub, Prague. 18 July 2002.

Kreuzmann, František. Actor: *The Sonnets*. Montmartre; Director: *Rozprava o Noci, Která se náhle stala Tajemnou*, Lyra Pragensis. 3 June 2002.

Kučera, Zdeněk. Producer: SSF. 28 August 2001.

Lang, Michal. Director: CD 94, Švandovo Divadlo. 27 August 2001.

Matásek, David. Actor. Graziano: *The Merchant of Venice*. Komedie. 2001; Petruchio: *The Taming of the Shrew*, Komedie; Albany: *King Lear*, SSF, 2002. 29 August 2001; 15, 25 July 2002.

Mejzlík, Miroslav. Actor. Antonio: *The Merchant of Venice*. Komedie. 28 August 2001.

Mrvová, Lenka. Actor. Viola, Captain: *Twelfth Night*; Irma the Feminist: *Úplněk*. Pidivadlo. 18 June 2002.

Pavlík, Tomáš. Director: *The Sonnets*. Montmartre. 3 June, 12 July 2002.
Peterková, Lucie. Actor. Clown: *Twelfth Night*; Theodora: *Úplněk*. Pidivadlo. 18 June 2002.
Pokorná, Zuzana. Actor. Olivia: *Twelfth Night*; Marie: *Úplněk*. Pidivadlo. 18, 19 June 2002.
Potměšil, Jan. Actor. Richard III: *Richard III*. Divadlo v Celetné. 26 July 2002.
Prachař, David. Actor. Portia: *The Merchant of Venice*. Komedie. 2001; Faust, Komedie, 2001; Clown: *King Lear*, SSF, 2002. 30 August 2001; 17 June 2002.
Přeučil, Jan. Actor. Lucencio's father: *The Taming of the Shrew*, SSF, 2000–1. *Love's Labours Lost*, SSF 2001–2. 29 August 2001.
Přibyl, Daniel. Translator and Shakespeare scholar. 26 July and 3 August 2001.
Rajmont, Ivan. Director: Národní Divadlo (National Theatre). 29 August 2001.
Rychlý, Michal. Producer: SSF. 21 and 28 August 2001.
Sitta, Martin. Actor. Biondello: *The Taming of the Shrew*, SSF, 2000–1; Tranio: *The Taming of the Shrew*, CD 94. 15, 17 and 28 August 2001.
Sloupová, Jitka. Translator and theatre critic, editor for *SAD* (*World and Theatre*). 30 August 2001.
Špalková, Petra. Actor. Katharina: *The Taming of the Shrew*, CD 94. 31 August 2001.
Štréblová, Alena. Actor. Katharina: *The Taming of the Shrew*, Komedie. 31 August 2001.
Stříbrný, Zdeněk. Shakespeare scholar, Professor Emeritus of English and American Literature at Charles University. 24 July and 4 August 2001.
Strnisko, Vladimír. Director: Slovenské Národní Divadlo (Slovak National Theatre); SSF. 9 August 2001.
Trtík, Rostislav. *The Sonnets*. Montmartre. 3 June 2002.

Scholarly publications

Abrams, Bradley F. *Struggle for the Soul of the Nation: Czech Culture and the Rise of Communism*. New York: Rowman and Littlefield, 2005.
Baranovsky, Vladimir. 'Russian Views on NATO and the EU', in Anatol Lieven and Dmitri Trenin (eds) *Ambivalent Neighbors: The EU, NATO and the Price of Membership*. Washington: Carnegie Endowment for International Peace, 2003: 269–94.
Bate, Jonathan. 'Shakespearean Nationhoods', in Michael Hattaway, Boika Sokolova and Derek Roper (eds) *Shakespeare in the New Europe*. Sheffield Academic Press, 1994: 112–29.
Bean, John C. 'Comic Structure and the Humanizing of Kate in *The Taming of the Shrew*', in Carolyn Ruth Swift Lenz *et al.* (eds) *The Woman's Part: Feminist Criticism of Shakespeare*. Urbana: University of Illinois Press, 1980: 65–78.
Bermann, Sandra and Michael Wood (eds). *Nation, Language, and the Ethics of Translation*. Princeton University Press, 2005.

Blumenfeld, Odette-Irenne. 'Shakespeare in Post-Revolutionary Romania: The Great Directors are Back Home', in Michael Hattaway, Boika Sokolova and Derek Roper (eds) *Shakespeare in the New Europe*. Sheffield Academic Press, 1994: 230–46.

Bobinski, Christopher. 'Polish Illusions and Reality', in Anatol Lieven and Dmitri Trenin (eds) *Ambivalent Neighbors: The EU, NATO and the Price of Membership*. Washington: Carnegie Endowment for International Peace, 2003: 231–44.

Boose, Lynda E. 'Scolding Brides and Bridling Scolds: Taming the Woman's Unruly Member', *Shakespeare Quarterly* 42, 2 (Spring 1992): 179–213.

_____ 'The Taming of the Shrew, Good Husbandry, and Enclosure', in Russ McDonald (ed.) *Shakespeare Reread: The Texts in New Contexts*. Ithaca: Cornell University Press, 1994: 193–225.

Bristol, Michael D. *Big-time Shakespeare*. London and New York: Routledge, 1996.

_____ 'Carnival and the Institution of Theater in Elizabethan England', *ELH* 50, 4 (Winter 1983): 637–54.

Carlson, Susan. 'The Suffrage Shrew: The Shakespeare Festival, "A Man's Play" and New Woman', in Jonathan Bate *et al.* (eds) *Shakespeare and the Twentieth Century*. Newark: Associated Press, 1998: 85–102.

Cartelli, Thomas: *Repositioning Shakespeare: National Formations, Postcolonial Appropriations*. London and New York: Routledge, 1999.

Černý, Ondřej and Vencl, Vít (eds). *Divadlo v České Republice 96–97*. Prague: Divadelní Ústav, 1998.

_____ *Divadlo v České Republice 97–98*. Prague: Divadelní Ústav, 1999.

_____ *Divadlo v České Republice 98–99*. Prague: Divadelní Ústav, 2000.

_____ *Divadlo v České Republice 99–00*. Prague: Divadelní Ústav, 2001.

_____ *Divadlo v České Republice 00–01*. Prague: Divadelní Ústav, 2002.

_____ *Divadlo v České Republice 01–02*. Prague: Divadelní Ústav, 2003.

_____ *Divadlo v České Republice 02–03*. Prague: Divadelní Ústav, 2004.

Černý, Ondřej and Jindrová, Zuzana (eds). *Divadlo v České Republice 95–96*. Prague: Divadelní Ústav, 1998.

Cerqueirová, Andrea. 'Partnerské soužití gayů a lesbiček: oříšek pro poslance', *Náš Svět* (August 2001): 242–4.

Chomsky, Noam. *Hegemony or Survival: America's Quest for Global Dominance*. New York: Metropolitan Books, 2003.

_____ *Interventions*. San Francisco: City Lights Books, 2007.

Christensen, Ann C. 'Petruchio's House in Postwar Suburbia: Reinventing the Domestic Woman (Again)', *Post Script* 17, 1 (Fall 1997): 28–42.

Cicero. 'Of Oratory', E.W. Sutton and H. Rackham (trans.), in Patricia Bizzell and Bruce Herzberg (eds) *The Rhetorical Tradition: Readings from Classical Times to the Present*. Boston: Bedford Books of St Martin's Press, 1990: 200–50.

Ciglar-Žanič, Janja. 'Recruiting the Bard: Onstage and Offstage Glimpses of Recent Shakespeare Productions in Croatia', in Michael Hattaway, Boika Sokolova and Derek Roper (eds) *Shakespeare in the New Europe*. Sheffield Academic Press, 1994: 261–75.

Čornej, Petr and Pokorný, Jiří. *Dějiny Českých Zemí do roku 2000 ve zkratce.* Prague: Práh, 2000.

Cronin, Michael. *Translation and Globalization.* London and New York: Routledge, 2003.

Deer, Harriet A. 'Untyping Stereotypes: *The Taming of the Shrew*', in Sara Munson Deats and Lagretta Tallent Lenker (eds) *The Aching Hearth: Family Violence in Life and Literature.* New York: Plenum, 1991: 63–78.

Derrida, Jaques. *Specters of Marx: The State of the Debt, the Work of Mourning, and the New International,* Peggy Kamuf (trans.). New York: Routledge, 1994.

Detmer, Emily. 'Civilizing Subordination: Domestic Violence and *The Taming of The Shrew*', *Shakespeare Quarterly* 48, 3 (Fall 1997): 273–94.

Dolan, Frances. *The Taming of the Shrew: Texts and Contexts.* Boston and New York: Bedford Books of St Martin's Press, 1996.

Dollimore, Jonathan: 'Shakespeare and Theory', in Ania Loomba and Martin Orkin (eds) *Post-Colonial Shakespeares.* London and New York: Routledge, 1998: 259–76.

Duggan, Lisa. *The Twilight of Equality? Neoliberalism, Cultural Politics, and the Attack on Democracy.* Boston: Beacon Press, 2003.

Egan, Gabriel. *Shakespeare and Marx.* Oxford University Press, 2004.

Fanel, Jiří. *Gay Historie.* Prague: Dauphin, 2002.

Freedman, Barbara. 'Frame-Up: Feminism, Psychoanalysis, Theatre', in Helene Keysar (ed.) *Feminist Theatre and Theory.* New York: St Martin's Press, 1996: 78–108.

Fujita, Minoru. 'Introduction', in Fujita Minoru and Leonard Pronko (eds) *Shakespeare East and* West. New York: St Martin's Press, 1996: 1–22.

Gal, Susan and Kligman, Gail. *The Politics of Gender After Socialism.* Princeton University Press, 2000.

Garner, Shirley Nelson. '*The Taming of the Shrew*: Inside or Outside the Joke?', in Maurice Charney (ed.) *'Bad' Shakespeare: Revaluations of the Shakespeare Canon.* Rutherford: Farleigh Dickinson University Press, 1988: 105–19.

Gay, Penny. 'Recent Australian Shrews: The "Larrikin Element"', in Marianne Novy (ed.) *Transforming Shakespeare: Contemporary Women's Revisions in Literature and Performance.* New York: St Martin's Press, 1999: 35–50.

Gerstle, Andrew: 'Shakespeare and Japanese Theatre: Artists' and Scholars' Use of the "Exotic"', in Fujita Minoru and Leonard Pronko (eds) *Shakespeare East and* West. New York: St Martin's Press, 1996: 61–76.

Giroux, Henry A. *The Terror of Neoliberalism: Authoritarianism and the Eclipse of Democracy.* Boulder: Paradigm, 2004.

Grabbe, Heather. 'Challenges of EU Enlargement', in Anatol Lieven and Dmitri Trenin (eds) *Ambivalent Neighbors: The EU, NATO and the Price of Membership.* Washington: Carnegie Endowment for International Peace, 2003: 67–89.

Guillory, John. *Cultural Capital: The Problem of Literary Canon Formation.* University of Chicago Press, 1994.

Habermas, Jürgen. *The Postnational Constellation: Political Essays,* Max Pensky (ed. and trans.). Cambridge, MA: MIT Press, 2001.

Hapgood, Robert. 'Popularizing Shakespeare: The Artistry of Franco Zeffirelli', in Linda Boose and Richard Burt (eds) *Shakespeare, the Movie: Popularizing the Plays on Film, TV, and Video*. London: Routledge, 1997: 80–94.
Harbage, Alfred. 'Shakespeare and the Myth of Perfection', in James G. McManaway (ed.) *Shakespeare 400: Essays by American Scholars on the Anniversary of the Poet's Birth*. New York, Chicago, San Francisco, Toronto and London: Holt, Reinehart and Winston, 1964: 1–10.
Harvey, David. *A Brief History of Neoliberalism*. New York: Oxford University Press, 2007.
Hattaway, Michael, Sokolova, Boika and Roper, Derek (eds). *Shakespeare in the New Europe*. Sheffield Academic Press, 1994.
Havel, Václav *et al*. *The Power of the Powerless*. New York: M.E. Sharpe, 1985.
Hawkes, Terrence. *Meaning by Shakespeare*. New York: Routledge, 1992.
_____ 'Introduction', in Terrence, Hawkes (ed.) *Alternative Shakespeares*, vol. 2. London: Routledge, 1996: 1–16.
_____ *Shakespeare in the Present*. New York: Routledge, 2002.
Healy, Thomas. 'Past and Present Shakespeares', in John Joughin and John Drakakis (eds) *Shakespeare and National Culture*. Manchester University Press, 1997: 206–32.
Heaney, Peter F. 'Petruchio's Horse: Equine and Household Management in *The Taming of the Shrew*', *Early Modern Literary Studies* 4, 1 (May 1998): 12.
Henderson, Diana. 'A Shrew for the Times', in Lynda Boose and Richard Burt (eds) *Shakespeare, the Movie: Popularizing the Plays on Film, TV and Video*. London: Routledge, 1997: 148–68.
Hilský, Martin. 'Shakespeare in Czech: An Essay in Cultural Semantics', in Michael Hattaway, Boika Sokolova and Derek Roper (eds) *Shakespeare in the New Europe*. Sheffield Academic Press, 1994: 150–8.
_____ '"Slova, slova, slova": Znovu k překladům *Hamleta*', *Svět a Divadlo* 11, 6 (2000): 179–83.
_____ '"Anebcokoli chcete": Proměny svátečního času', in *William Shakespeare: Večer tříkrálový aneb cokoli chcete*. Prague: Národní divadlo, 2001: 7–26.
_____ 'Chvála bláznovství aneb zrod shakespearovského šaška', in *William Shakespeare: Večer tříkrálový aneb cokoli chcete*. Prague: Národní divadlo, 2001: 28–41.
_____ 'Umění harmonie a gramatika touhy', in *William Shakespeare: Večer tříkrálový aneb cokoli chcete*. Prague: Národní divadlo, 2001: 43–50.
_____ 'Z těch zrad jsem ale poznal, co je ryzí: České překlady Williama Shakespeara a zrada národní kultury', *Lidové Noviny*, 10 May 2001: 20.
Hodek, Břetislav. 'Neveselá 'Komedie Chyb': V diskusi nad překlady Shakespearova díla odpovídá Břetislav Hodek Martinu Hilskému', *Lidové Noviny*, 7 June 2001.
_____ 'Tečka za diskusí Anglistů', *Lidové Noviny* 14, 172 (July 2001): 28.
Hogan, Patrick Colm: 'Shakespeare, Eastern Theatre, and Literary Universals: Drama in the Context of Cognitive Science', in Fujita Minoru and Leonard Pronko (eds) *Shakespeare East and West*. New York: St Martin's Press, 1996: 164–80.

Holderness, Graham. *Cultural Shakespeare: Essays in the Shakespeare Myth*. University of Hertfordshire Press, 2001.

_____ *Visual Shakespeare: Essays in Film and Television*. University of Hertfordshire Press, 2002.

Horálek, Karel. 'The Folk Sources of Hamlet', in Zdeněk Stříbrný (ed.) *The Charles University on Shakespeare*. Prague: Univerzita Karlova, 1966: 115–25.

Howard, Jean E. and O'Connor, Marion F. 'Introduction', in Jean E. Howard and Marion F. O'Connor (eds) *Shakespeare Reproduced: The Text in History and Ideology*. New York and London: Methuen, 1987: 1–17.

Janošová, Pavlína. *Homosexualita v České Republice*. Prague: Karolina, 2002.

Jelínek, František. *Homosexualita ve světě vědy*. Prague: Obelisk, 1924.

Kamp, Karl-Heinz. 'The Dynamics of NATO Enlargement', in Anatol Lieven and Dmitri Trenin (eds) *Ambivalent Neighbors: The EU, NATO and the Price of Membership*. Washington: Carnegie Endowment for International Peace, 2003: 185–204.

Kennedy, Gwynne. 'Lessons of the "Schoole of Wisedome"', in Carole Levin and Karen Robertson (eds) *Sexuality and Politics in Renaissance Drama*. New York: Mellen Lewisnton, 1991: 113–36.

Klein, Naomi. *The Shock Doctrine: The Rise of Disaster Capitalism*. New York: Metropolitan Books, 2007.

Knowles, Richard. 'Focus, Faithfulness, Shakespeare, and the Shrew: Directing as Translation as Resistance', *Essays in Theatre* 16, 1 (November 1997): 33–52.

Kopecký Jan. 'Shakespeare's Forgotten Theatre: A Contribution to the Problems of the Theatre Today', in Zdeněk Stříbrný (ed.) *The Charles University on Shakespeare*. Prague: Univerzita Karlova, 1966: 93–113.

Korda, Natasha. 'Household Kates: Domesticating Commodities in *The Taming of the Shrew*', *Shakespeare Quarterly* 47, 2 (Summer 1996): 109–31.

Kostihová, Marcela. 'Katherina "Humanized": Abusing the "Shrew" on Prague Stages', in *Worldwide Shakespeares: Local Appropriations in Film and Performance*. New York and London: Routledge, 2005: 72–9.

Kott, Jan. *Shakespeare our Contemporary*. New York: Methuen, 1967.

Kraus, Karel. *Divadlo ve službách dramatu: Eseje, kritiky, analýzy*. Prague: Divadelní Ústav, 2001.

Kritzer, Amelia Howe. *Political Theatre in Post-Thatcher Britain: New Writing: 1995–2005*. Basingstoke: Palgrave Macmillan, 2008.

Kundera, Milan. 'The Tragedy of Central Europe', *The New York Review of Books* 31, 7 (26 April 1984): 33–8.

_____ 'Die Weltliteratur: How We Read One Another', *The New Yorker* (8 January 2007): 28–35.

Lanier, Douglas. *Shakespeare and Modern Popular Culture*. Oxford University Press, 2002.

Lewis, Cynthia: *Particular Saints: Shakespeare's Four Antonios, their Contexts, and their Plays*. Newark: Delaware University Press, London: Associate University Press, 1997.

Lieven, Anatol. 'Conclusions: The Pangs of Disappointed Love? A Divided West and its Multiple Peripheries', in Anatol Lieven and Dmitri Trenin (eds)

Ambivalent Neighbors: The EU, NATO and the Price of Membership. Washington: Carnegie Endowment for International Peace, 2003: 295–312.

Londré, Felicia Hardison. 'Confronting Shakespeare's "Political Incorrectness" in Production: Contemporary American Audiences and the New "Problem Plays"', *On-stage Studies* 19 (1996): 67–82.

Loomba, Ania and Orkin, Martin (eds). *Post-Colonial Shakespeares*. London and New York: Routledge, 1998.

Maguire, Laurie E. 'Cultural Control in *The Taming of the Shrew*', *Renaissance Drama* 26 (1995): 83–104.

Makaryk, Irena R. and Price, Joseph G. (eds). *Shakespeare in the Worlds of Communism and Socialism*. University of Toronto Press, 2006.

Malý, Jakub. *Shakespear a jeho díla*. Prague: Muzeum Království Českého, 1873.

Mathesius, Vilem. *Co Daly Naše Země Evropě a Lidstvu,* vols. 1–3, 2nd edn. Prague: Evropský literární klub, 1999.

McKee, Martin. 'Substance Use and Social and Economic Transition: The Need for Evidence', *The International Journal on Drug Policy* 13, 6 (2002): 453–7.

Milanovic, Branko. *Income, Inequality, and Poverty during the Transition from Planned to Market Economy*. Washington: World Bank, 1998.

Moisan, Thomas. '"Knock Me Here Soundly": Comic Misprision and Class Consciousness in Shakespeare', *Shakespeare Quarterly* 42, 3 (Fall 1991): 276–90.

Möller-Leimkühler, Anne Maria. 'The Gender Gap in Suicide and Premature Death or: Why are Men so Vulnerable?', *European Archive of Psychiatry and Clinical Neuroscience* 253 (2003): 1–8.

Moskalewicz, Jacek and Simpura, Jussi. 'Alcohol Supply in Developing and Transitional Societies: The Supply of Alcoholic Beverages in Transitional Conditions: The Case of Central and Eastern Europe', *Addiction* 95, 4 (2000): 505–22.

Moskalewicz, Jacek. 'Alcohol in the Countries in Transition: The Polish Experience and the Wider Context', *Contemporary Drug Problems* 27 (Fall 2000): 561–92.

Motyl, Alexander J. 'Ukraine, Europe and Russia: Exclusion or Dependence?', in Anatol Lieven and Dmitri Trenin (eds) *Ambivalent Neighbors: The EU, NATO and the Price of Membership*. Washington: Carnegie Endowment for International Peace, 2003: 15–43.

Mukařovský, Jan: 'Shakespeare and Czech Theatrical Criticism', in Zdeněk Stříbrný (ed.) *The Charles University on Shakespeare*. Prague: Univerzita Karlova, 1966: 11–23.

Musial, Petr. 'Úvod', *Postavení homosexuálův ČSR v letech 1918–1938*. Published as a series in *Náš Svět*. September 2001: 310–12.

Musilová, Martina. 'Krotitelé', *SAD* 6 (2000): 50–3.

Neruda, Jan. *Kzákladům Národního Divadla,* Adolf Wenig (ed.). Prague: F. Topič, Knihkupec, 1918.

______ *České divadlo,* vols. 1–6. Prague: Odeon, Knihovna Klasiků, 1966.

Neumayer, Laure. 'Euroscepticism as a Political Label', *European Journal of Political Research* 47 (2008): 135–60.

Oates-Indruchová, Libora. 'The Void of Acceptable Masculinity During Czech State Socialism: The Case of Radek John's *Memento*', *Men and Masculinities* 8, 4 (April 2006): 428–50.

Okey, Robin. *The Demise of Communist East Europe: 1989 in Context*. London: Oxford University Press, 2004.

Orgel, Stephen. *Impersonations: The Performance of Gender in Shakespeare's England*. Cambridge University Press, 1996.

Orkin, Martin. 'Shifting Shakespeare', *PMLA* 118, 1 (January 2003): 134–6.

_____ *Local Shakespeares: Proximations and Power*. London: Routledge, 2005.

Ostrovsky, Arcady. 'Shakespeare as a Founding Father of Socialist Realism: The Soviet Affair with Shakespeare', in *Shakespeare in the Worlds of Communism and Socialism*. University of Toronto Press, 2006: 56–83.

Otáhal, Milan. *Podíl tvůrčí inteligence na pádu komunismu: Kruh nezávislé inteligence*. Brno: Doplněk, 1999.

Ozolina, Zaneta. 'The EU and the Baltic States', in Anatol Lieven and Dmitri Trenin (eds) *Ambivalent Neighbors: The EU, NATO and the Price of Membership*. Washington: Carnegie Endowment for International Peace, 2003: 205–30.

Pancheva, Evgenia. 'Nothings, Merchants, Tempests: Trimming Shakespeare for the 1992 Bulgarian Stage', in Michael Hattaway, Boika Sokolova and Derek Roper (eds) *Shakespeare in the New Europe*. Sheffield Academic Press, 1994: 247–60.

Patel, Tara. 'Alcohol Takes its Toll on Post-communist Europe', *New Scientist*, 23 (December 1995): 77.

Pečírka, Jaromír: 'Shakespeare and the Graphic Arts', in Zdeněk Stříbrný (ed.) *The Charles University on Shakespeare*. Prague: Univerzita Karlova, 1966: 83–92.

Pfister, Manfred. 'Hamlets Made in Germany, East and West', in Michael Hattaway, Boika Sokolova and Derek Roper (eds) *Shakespeare in the New Europe*. Sheffield Academic Press, 1994: 76–91.

Polišenský, Josef. 'England and Bohemia in Shakespeare's day', in Zdeněk Stříbrný (ed.) *The Charles University on Shakespeare*. Prague: Univerzita Karlova, 1966: 65–81.

Přibyl, Daniel. (2001a) 'Anglická inscenační tradice', in *William Shakespeare: Večer tříkrálový aneb cokoli chcete*. Prague: Národní divadlo, 2001: 56–65.

_____ (2001b) 'Večer tříkrálový v Národním divadle', in *William Shakespeare: Večer tříkrálový aneb cokoli chcete*. Prague: Národní divadlo, 2001: 67–86.

Quince, Rohan. *Shakespeare in South Africa: Stage Productions During the Apartheid Era*. Studies in Shakespeare, vol. 9, Robert F. Willson (gen. ed.). New York: Peter Lang Publishing, 2000.

Ranald, Margaret Loftus. '"As Marriage Binds, and Blood Breaks": English Marriage and Shakespeare,' *Shakespeare Quarterly* 30 (1979): 68–81.

Roberts, Jeanne Addison. 'Horses and Hermaphrodites: Metamorphoses in *The Taming of the Shrew*', *Shakespeare Quarterly* 34, 2 (Summer 1983): 159–71.

Said, Edward W. *Culture and Imperialism*. New York: Vintage, 1994.

Scribner, Charity. *Requiem for Communism*. Cambridge, MA: MIT Press, 2003.

Shakespeare v Čechách. Documentary. Directed by Ondřej Šrámek, moderator Václav Helšus, Česká Televize, 2001.

Shakespeare, William. Complete plays. Editions used: Stephen Greenblatt *et al.* (eds). *The Norton Shakespeare*. New York: Norton, 1997; and *The Riverside Shakespeare*, 2nd edn. Boston and New York: Houghton Mifflin, 1997.

Shrubanov, Alexander and Sokolova, Boika. 'From the Unlove of *Romeo and Juliet* to *Hamlet* Without the Prince: A Shakespearean Mirror Held up to the Fortunes of New Bulgaria', in Michael Hattaway, Boika Sokolova and Derek Roper (eds) *Shakespeare in the New Europe*. Sheffield Academic Press, 1994: 24–53.

Siberry, Michael. 'Petruchio in *The Taming of the Shrew*', in Robert Smallwood (ed.) *Players of Shakespeare 4: Further Essays in Shakespearean Performance by Players with the Royal Shakespeare Company*. Cambridge University Press, 1998: 45–59.

Simpura, Jussi. 'Alcohol in Eastern Europe: Market Prospects, Prevention Puzzles', *Addiction* 90 (1995): 467–70.

Sinfield, Alan. 'Introduction: Reproductions, Interventions', in Jonathan Dollimore and Alan Sinfield (eds) *Political Shakespeare: Essays in Cultural Materialism*. Ithaca: Cornell University Press, 1994: 154–7.

Škapíková, Jitka. 'Shakespearův Hilský, Hilského Shakespeare: Kulturní diskuse', *Lidové Noviny*, 21 June 2001: 28.

______ 'Někdy mám pocit, že zastupuji samotného Shakespeara: Rozhovor s anglistou a překladatelem Martinem Hilským', *Týdeník Rozhlas* 27 (2–8 July 2001): 1, 5.

Škvorecký, Josef. 'To the Editors', *The New York Review of Books* 33, 7 (24 April 1986): 49.

Slach, Miroslav. *Usměvaví Rebelanti*. Prague: Melantrich, 1959.

Slomek, Jaromír. 'Co komu Shakespeare udělal? Ptá se nad současným zacházením s klasikovými texty anglista a shakespearolog Břetislav Hodek', *Lidové Noviny*, 26 April 2001: 20.

Sloupová, Jitka. 'Co Čech, to Hamlet: Enjoying Hamlet', *Svět a Divadlo* 4 (2000): 8–25.

______ 'Slova, na nichž trvám', *Svět a Divadlo* 1 (2001): 161–3.

Sokolyansky, Mark. '"Giant-like Rebellions" and Recent Russian Experience: Shakespearean Irony as an Approach to Modern History', in Michael Hattaway, Boika Sokolova and Derek Roper (eds) *Shakespeare in the New Europe*. Sheffield Academic Press, 1994: 221–9.

Stallybrass, Peter. 'Editing as Cultural Formation: The Sexing of Shakespeare's *Sonnets*', in Stephen Orgel and Sean Keilen (eds) *Shakespeare's Poems*. New York: Garland Publishing, 1999: 133–45.

Štěpánek, Vladimír. 'The Importance of Shakespeare for the Formation of Modern Czech Literature', in Zdeněk Stříbrný (ed.) *The Charles University on Shakespeare*. Prague: Univerzita Karlova, 1966: 127–42.

Stříbrný, Zdeněk (ed.). *The Charles University on Shakespeare*. Prague: Univerzita Karlova, 1966.

______ 'Shakespeare Today', in Zdeněk Stříbrný (ed.) *The Charles University on Shakespeare*. Prague: Univerzita Karlova, 1966: 25–37.

_____ *Shakespeare and Eastern Europe.* Oxford University Press, 2000.

_____ *The Whirligig of Time: Essays on Shakespeare and Czechoslovakia,* Lois Potter (ed.). Newark: University of Delaware Press, 2007.

Štulhofer, Aleksander and Sandfort, Theo. 'Introduction: Sexuality and Gender in Times of Transition', in *Sexuality and Gender in Postcommunist Eastern Europe and Russia.* New York: The Haworth Press, 2005: 1–25.

Swain, Adam. 'Soft Capitalism and Hard Industry: Virtualism, the "Transition Industry" and the Restructuring of the Ukrainian Coal Industry', *Transactions of the Institute of British Geographers* 31, 2 (June 2006): 208–23.

Szacki, Jerzy. *Liberalism After Communism.* Budapest and New York: Central European University Press, 1996.

Trenin, Dmitri. 'Introduction: The Grand Redesign', in Anatol Lieven and Dmitri Trenin (eds) *Ambivalent Neighbors: The EU, NATO and the Price of Membership.* Washington: Carnegie Endowment for International Peace, 2003: 1–14.

Trnka, Bohumil. 'Shakespeare's Ethics and Philosophy', in Zdeněk Stříbrný (ed.) *The Charles University on Shakespeare.* Prague: Univerzita Karlova, 1966: 55–63.

True, Jacqui. *Gender, Globalization, and Postsocialism: The Czech Republic After Communism.* New York: Columbia, 2003.

Venuti, Lawrence. *The Translators Invisibility: A History of Translation,* 2nd edn. London and New York: Routledge, 2008.

Verdery, Katherine. *What was Socialism, and What Comes Next?* Princeton University Press, 1996.

Wallace, William. 'Does the EU have an *Ostpolitik*?', in Anatol Lieven and Dmitri Trenin (eds) *Ambivalent Neighbors: The EU, NATO and the Price of Membership.* Washington: Carnegie Endowment for International Peace, 2003: 44–66.

Wedel, Janine R. *Collision and Collusion: The Strange Case of Western Aid to Eastern Europe, 1989–1998.* New York: St Martin's Press, 1998.

Wejnert, Barbara (ed.). *Transition to Democracy in Eastern Europe and Russia.* Westport: Praeger, 2002.

Zdeněk, Vančura. 'Shakespeare – Whose Contemporary?', in Zdeněk Stříbrný (ed.) *The Charles University on Shakespeare.* Prague: Univerzita Karlova, 1966: 39–54.

Newspaper articles/performance reviews

AP. 'Násilí na lidech i zvířatech spolu souvisí', *Mladá Fronta Dnes,* 4 July 2001.

avo and tof. 'Domácí násilí většina lidí na Vysočině vůbec nezná', *Mladá Fronta Dnes,* 14 November 2001.

Bártová, Milena. 'Klec bláznů nabízí únik z reality: Heterosexuál Ladislav Županič exceluje v roli homosexuála', *Blesk,* 14 January 1999.

Bělíková, Markéta. 'Násilí partnera nelze omlouvat', *Mladá Fronta Dnes,* 18 December 2000.

Círtek, Pavel. 'O ženě, která by měla být zkrocená', *Právo*, October 1997: 12.
čtk. 'Oběti násilí postrádají pomoc', *Mladá Fronta Dnes*, 6 December 2000.
_____ 'Nová rada vlády bude dohlížet na práva žen: Projekt předpokládá vyškolení asi 300 úředníků', *Lidové Noviny*, 10 October 2001.
_____ 'Shakesparovy slavnosti déšt nesplachl', *Hospodářské Noviny*, 18 August 2005: 18.
dam. 'Domácího násilí neustále přibývá', *Mladá Fronta Dnes*, 25 September 2001.
Daníčková, Sylva. 'Černá dáma Sonetů', *Svobodné slovo*, 14 August 1986.
dík. 'Domácí násilí se většinou nesoudí', *Mladá Fronta Dnes*, 11 October 2001.
Dolanský, Lukáš. (2001a) 'Nejsme Připraveni', *Mladá Fronta Dnes*, 23 May 2001.
_____ (2001b) 'Svědci si obvykle ničeho nevšímají', *Mladá Fronta Dnes*, 23 May 2001.
Erml, Richard. (2000a) 'Totálně zkrocená Kateřina', *Reflex* 40 (2000): 70.
_____ (2000b) 'Jak hrát Sylocka po Holocaustu?', *Divadelní Noviny* 9, 9 (2 May 2000): 2.
Frajerová, Blanka. 'Županič doufá, že diváci budou hodně tolerantní: Nový karlínský muzikál má být hlubokým lidským příběhem', *Zemské Noviny*, 28 May 1998: 56.
_____ 'Česká Klec bláznů se vyvedla: Ladislav Županič dal do svého Albína všechen svůj herecký talent a cit', *Zemské Noviny*, 1 June 1998.
Francková, Illona. 'Dočekalovo Zkrocení sází plně na divákovu vnímavost', *Zemské Noviny*, 9 September 2000: 11.
Guryča, Richard. 'Domácí násilí ničí tisíce žen', *Mladá Fronta Dnes*, 3 February 2001.
Herman, Josef. 'Divadlo: Muzikál: Jsme to, co jsme!', *Týden*, 15 June 1998.
_____ 'Norma: "Miluji Francouze!" Toddy: "Já také"', *Divadelní Noviny* 9, 22 (27 November 2000): 5.
Hrbotický, Saša. 'Večer Tříkrálový humor ani tempo nepostrádá', *Hospodářské Noviny*, 25 July 2005: 7.
Hrdinová, Radmila. 'V Karlíně se zkouší Klec Bláznů', *Právo*, 13 May 1998: 10.
_____ 'V Karlíně je Klec na intolerance', *Právo*, 5 June 1998: 13.
_____ 'Shakespearovský duel za zvuku bubnů', *Právo*, 7 April 2000: 12.
_____ 'Karlín chystá transvestitní muzikál', *Právo*, 21 November 2000: 15.
_____ 'Když žena hraje chlapa, který hraje ženu', *Právo*, 29 November 2000: 12.
Hulec, Vladimír. 'Shakespearovy Sonety plynou v předvídatelném toku', *Mladá Fronta Dnes*, 2 February 1996: 18.
jam. 'Studio natočilo film o násilí', *Mladá Fronta Dnes*, 25 November 2001.
Jeníková, Eva. (2000a) 'Shakespeare na Pražském Hradě bude Slovensko-Český, erotický, i němý', *Zemské Noviny*, 15 June 2000.
_____ (2000b) 'Zlá žena zaráží vulgaritami', *Zemské Noviny*, 3 July 2000.
_____ (2000c) 'Dolinová se předvede v převlekové roli', *Zemské Noviny*, 24 November 2000: 10.
-jp. 'Večer slova a hudby', *Večerní Praha*, 21 August 1975.

jš. 'Shakespeare nepřeje dovoleným', *Nedělní Svět* 11:27, 3 July 2005: 1.
Jungmanová, Lenka. 'Divadla: Kam za divadlem', *Přehled kulturních pořadů v Praze* 11 (1998): 9–10.
_____ 'Večeru Tříkrálovému v ABC schází pořádná akce', *Lidové Noviny*, 19 February 1999.
Just, Vladimír. 'Úvahy postdivadelní: Vitězství karnevalu', *Literární Noviny*, 3 March 1999.
Kašáková, Eva. 'Shakespearovo poselství k divákům i po několika stoletích', *Zemské noviny*, 11 April 1996: 14.
Kerbr, Jan. 'Nejvíc jsem vyřezával, když jsem měl málo práce', *Divadelní Noviny* 10, 7 (12 May 1998): 1, 6–7.
_____ 'Zase krotíme Kateřinu', *Mladá Fronta Dnes*, 31 October 2000.
_____ 'Večer Tříkrálový: Plenérový standard s rozpačitým rozjezdem', *Reflex* 16:29 (21 July 2005): 20.
Kolář, Jan. 'Hravý, dravý, ale i nevyrovnaný je Večer Tříkrálový', *Mladá Fronta Dnes*, 24 February 1999.
Kolářová, Kateřina. 'V Komedii uvedou Zkrocení i s epilogem', *Mladá Fronta Dnes*, 14 September 2000.
Kolářová, Lucie. 'Násilí v rodině je trestným činem', *Mladá Fronta Dnes*, 12 April 2001.
'Kritický žebříček', *Divadelní Noviny* 18 (25 May 2001): 3.
-ku. 'Shakespearovy Sonety', *Lidová Demokracie*, 19 January 1991: 4.
lds. 'Násilí zná z domova každý šestý', *Mladá Fronta Dnes*, 23 May 2001.
Lukeš, Alexander. 'Cikána za souseda? Nikdy!: Češi by Romy poslali do školy, do práce, ale i do rezervace', *Týden* 21 (20 May 2002): 30–1.
Lukeš, Milan. 'Podnikatelé benátští', *Divadelní Noviny* 6, 13 (24 June 1997): 5.
Machalická, Jana. 'Zrcadlová hra stínů a světel: Nový překlad Shakespearových Sonetů v Labyrintu', *Lidové Noviny*, 24 April 1997: 11.
_____ 'Zkrocení zlé ženy jako uspěchaná fraška', *Lidové Noviny*, 3 July 2000.
-metro. 'Viktor i Viktorie řádí v Karlíně', *Metro*, 15 December 2000: 11.
Mikulka, Vladimír. 'Trabant, Koubek, Mafiáni a nastudování jednoho Shakespeara', *Divadelní Noviny* 16 (2000): 6, 23.
_____ 'Aneb cokoliv nás napadlo', *Divadelní Noviny* 17 (2001): 5.
_____ 'Lidem se to líbí – a hotovínko', *Kritická příloha k Revolver Revue*' 22 (2002): 25–38.
Mlejnek, Josef. 'Přízemnost versus láska ve Večeru Tříkrálovém', *Mladá Fronta Dnes*, 27 June 2005: 3–4.
Paterová, Jana. 'Večer plný omylů a lásky: Letní shakespearovské slavnosti 2005 zahájily komedií Večer tříkrálový', *Lidové Noviny*, 27 June 2005.
Pawlicová, Marta. 'Rosnička v roli ženochlapa', *Kladenský deník*, 9 January 2001: 14.
pdk. 'Domácí násilí zná nejméně 16 procent Čechů', *Lidové Noviny*, 7 July 2001.
Pekárková, Milena. 'Muzikál Klec Bláznů', *Super*, 8 June 2001: 13.
Polcarová, Simona. 'Polesný vsadil na legraci', *Rovnost*, 26 July 2005: 15.

Pražan, Bronislav. 'Divadlo Komedie: Kupec Benátský 2001', *Rozhlas* 25 (1997): 4.

Prchalová, Radka. 'Divadlo ABC se Večerem Tříkrálovým na čas loučí', *Mladá Fronta Dnes*, 12 February 1999.

Reslová, Marie. 'O kupeckých počtech, milosti a právu: Shakespearův Kupec benátský jako fikce na téma xenofobie', *Lidové Noviny* 10, 122 (27 May 1997): 13.

rh. 'Shakespearovské slavnosti už zkouší Večer Tříkrálový', *Právo*, 21 April 2005: 11.

Rogl, Vladimír. 'Michaela Dolinová ve fraku – holka nebo kluk?', *Večerník Praha: Společnost*, 21 November 2000: 15.

Rohlíček, Zdeněk. 'V novém karlínském muzikálu všechno hraje ve velkém stylu', *Svobodné Slovo: Kultura*, 16 June 1998: 9.

-rt. 'Zrcadlo nastavené životu', *Slovo*, 21 May 1993: 3.

'Shakespeare opět vládne na Pražském Hradě', *Time In* 3, 7 (30 June 2005): 60–6.

sib. 'Pojem, domácí násilí' má být v zákoně', *Lidové Noviny*, 28 October 2001, reprinted 21 November 2001.

Sloupová, Jitka. 'Divadlo bez oddechu', *Divadelní Noviny* 9, 21 (12 December 2000): 2.

Soprová, Jana. 'Shakespeare psal také básně: o životě, lásce, přátelství – o vás pro vás', *Večerní Praha*, 29 May 1996: 13.

______ 'Jak se krotí d'áblice: Shakespearova hra pro Antifeministy', *Večerník Praha*, 1 December 1997.

______ 'Klec bláznů: Česká premiéra poněkud šokujícího muzikálu', *Večerník Praha*, 29 May 1998.

______ 'V Kleci bláznů s Albínou: Česká premiéra zbrusu nového muzikálu v Hudebním divadle v Karlíně', *Večerník Praha*, 19 June 1998.

'Soukup a Županič ztvární dvojici homosexuálů: Českou premiéru si americký muzikál Klec Bláznů odbude na konci května v Karlíně', *Boleslavský deník*, 22 May 1998.

Stanislavčík, Tomáš. 'Ladislav Županič hraje transvestitu a vyzývá k toleranci', *Lidové Noviny*, 1 May 1998: 1.

šv. 'Divadlo ABC v Praze: Poslední Premiéra před rekonstrukcí', *Týden*, 1 March 1999: 10.

sve. 'Násilníci napadají nejen svoje družky a manželky, ale i matky', *Mladá Fronta Dnes*, 21 February 2001.

Šrámková, Vítězslava. 'Vše co je krásné, má se rozmnožovat', *Prapor* 52 (1996): 13.

______ 'Kruté hry spolu lidé hrají', *Týdeník Rozhlas* 44 (27 October–2 November 1997): 4.

Švagrová, Marta. 'Klec bláznů: Aktuální premiéra v Karlíně', *Týden*, 25 May 1998.

Šverdík, Michal. 'Oběti domácího násilí maří trest', *Mladá Fronta Dnes*, 12 December 2001.

Tesárková, Radka. 'Shakespearovy Sonety na Scéně Studia Labyrint', *Svobodné Slovo*, 30 April 1996: 8.

_____ 'Režisér Rupert Dubský: Hudební svět mi připadá vyváženější', *Slovo,* 30 May 1998: 10.

_____ '*Večer Tříkrálový* zraje jako víno: Režisér Milan Shejbal si vyzkoušel herecký chlebíček', *Večerní Praha,* 17 June 1999.

Tichý, Zdeněk A. 'Krotit Špalkovou se vyplácí', *Mladá Fronta Dnes,* 10 October 1997.

_____ 'Karlín uvádí muzikál Klec bláznů', *Mladá Fronta Dnes,* 29 May 1998.

_____ 'Vstoupit do muzikálové klece bláznů není nebezpečné', *Mladá Fronta Dnes,* 23 June 1998.

_____ 'Karlín uvede oscarový hit', *Mladá Fronta Dnes,* 25 November 2000: 19.

-tof. 'My jsme to co jsme', *Hospodářské noviny: kultura,* 12 June 1998: 46.

Turková, Martina. 'Splnil se jí sen být mužem', *Hradecké noviny: Víkendové čtení,* 24 February 2001: 10.

Ungermann, Libor. 'Asi byste mě nezaplatili' (interview with Jane Elliott, choreographer of *Victor-Viktoria*) *Reflex* 11, 49 (7 December 2000): 75.

Urbanová, Alena. (1997a) 'Rozverné hravé krocení zlé ženy v Činoherním Klubu', *Denní Telegraf* 6, 234 (6 October 1997): 13.

_____ (1997b) 'Rozpolcený Kupec Benátský v divadle Komedie', *Denní Telegraf,* 30 May 1997: 11.

'Viktor Polesný: Režírovat Shakespeara je velké dobrodružství', *Time In,* 3:7 (30 June 2005): 78–80.

Vlastník, Jiří. 'Česko-Slovenské krocení zlé ženy', *Kladenský Denník,* 25 May 2000.

-vš. 'Láska je horečka a rozum je jejím lékařem, praví Shakespeare', *Zemské Noviny,* 20 June 1996: 14.

Závodský, Vít. 'Kam se ubírá tradiční opereta a průbojný muzikál?', *Kam* 7, 12 (December 2001): 2–4.

Zídek, Petr. 'Diskriminováni jsou muži', *Lidové Noviny,* 10 October 2001.

Referenced performances (listing by directors)

Dočekal, Michal. William Shakespeare: *Kupec Benátský (The Merchant of Venice).* Komedie. 1997. Also available on video at Divadelní Ústav archive.

_____ William Shakespeare: *Král Lear* (*King Lear*). Divadlo Komedie, 2001.

_____ William Shakespeare: *Bouře.* (*The Tempest*). Divadlo Komedie, 2002.

_____ Christopher Marlowe: *Faustus.* Divadlo Komedie, Gorlice. 2001.

_____ William Shakespeare: *Zkrocení Zlé Ženy* (*The Taming of the Shrew*). Divadlo Komedie, also available on video at Divadelní Ústav archive.

Dubský, Robert. Henry Mancini and Blake Edwards: *Viktor-Viktorie (Victor-Victoria).* Hudební Divadlo Karlín, 15 June 2002.

Ezseny, Enikő. William Shakespeare: *Večer Tříkrálový* (*Twelfth Nitght*). Stavovské Divadlo (a stage of the National Theatre), 24 June 2002.

Herz, Juraj and Ephraim Kishon. *Byl to pták: Romeo a Julie po třiceti letech.* (*It was a Lark: Romeo and Juliet Thirty Years Later*). Divadlo bez Zábradlí, 2000.

Hlaučo, Miroslav. Miroslav Hlaučo *et al.*: *Úplněk (The Full Moon).* Pidivadlo (VOŠ Herecká), 13 June 2002.

_____ William Shakespeare: *Večer Tříkrálový* (*Twelfth Nitght*). Pidivadlo (VOŠ Herecká), 28 May, 18 June 2002.

Huba, Martin. William Shakespeare: *Král Lear* (*King Lear*). Letní Shakespearovské Slavnosti, 2002.

Kreuzmann, František. Tommaso Landolfi's: *Rozprava o Noci, Která se náhle stala Tajemnou*. Lyra Pragensis, 2002.

Lang, Michal. William Shakespeare: *Zkrocení Zlé Ženy* (*The Taming of the Shrew*). Činoherní Klub, CD 94, 1997.

_____ William Shakespeare: *Bouře* (*The Tempest*). CD 94, 2000.

Pavlík, Tomáš. William Shakespeare: *Sonety (The Sonnets)*. Divadlo Montmartre (Divadlo ve Městě), 9 June 2002.

Polesný, Viktor. *Večer Tříkrálový* (*Twelfth Night*). Letní Shakespearovske Slavnosti (Summer Shakespeare Festival), June–August 2005.

Rajmont, Ivan. William Shakespeare: *Cymbelín* (*Cymbeline*). Národní Divadlo, 1997.

Schejbal, Milan. *Večer Tříkrálový* (*Twelfth Night*). Divadlo ABC. 1999–2006.

Špalek, Jakub. William Shakespeare: *Richard III*. Divadlo v Celetné, 2000.

Index